YOUR MEXICAN FLIGHT PLAN

YOUR MEXICAN FLIGHT PLAN

BY DON & JULIA DOWNIE

FIRST EDITION

FIRST PRINTING

Printed in the United States of America

Library of Congress Cataloging in Publication Data

Downie, Don.
Your Mexican flight plan.

Includes index.
1. Air-pilot guides—Mexico. I. Downie, Julia.
II. Title.
TL726.5.M4D68 1983 917.2'04'0246291 82-19392
ISBN 0-8306-2337-X (pbk.)

Contents

Introduction

If you want to visit a different culture, exotic beaches, ancient pyramids and isolated villages wherein people live as they did a century ago; if you want to challenge the newness of a foreign neighbor with different rules and regulations and a pace of living and customs unlike those of the United States; if you want to enjoy the mild adventure and a view of tropics and snowcapped peaks different from those with which you've grown up, then you'll look forward to flying yourself to Mexico.

For all practical purposes, you can forget the horror stories of years past in which bribes were the order of the day and tourists were detained and their aircraft confiscated for an unwitting minor infraction of local law. However, there are always people and circumstances that can lead to problems—the same ones you'd get on either side of the Border.

Approach your personal flight to Mexico by remembering that you, your passengers, and your aircraft are guests in a neighboring country. Conform with *their* rules, *their* procedures and *their* pace of activity and you should have no real problems.

There will be times when the "mañana syndrome" will start to drive you up the walls. Take a border crossing stop where it's three hours to dark and that destination you've planned on for months is just two hours and forty-five minutes flying time away, and you can't find the airport official required to sign that last piece of paper, or the man running the gas pump has gone to town and will be back "in about 30 minutes." It's frustrating, of course, but the only thing you can do is "cool it," go along with the Mexican pace. If you don't actually get to your preplanned destination that night, there's always "mañana."

The Mexican government has done a fine, conscientious job of improving conditions for the tourist, whether on the ground or in the air. Every procedure around Mexican airports has been improved. Way back when, there were many problems that influenced visitors never to return. When you compare today's conditions with those of 15 years ago, it's absolutely great!

No matter where you live in the U.S., Mexico isn't all that far when you do your own flying. A visit south of the border is a complete contrast to what you may see in Canada or the States. Mexico has a completely different culture, in many cases a race apart where millions of Najuas and Mixtecs, Mayans or Coras have Spanish as a second language.

Tourism is an important and established part of Mexico's economy, and the smiles you receive are sincere. They may be tainted just slightly by the number of pesos you will spend in Mexico, but on the other hand, remember that you're getting a bargain when you compare travel costs on other continents.

Don't expect the same kind of blistering efficiency that sometimes develops north of the border. With the exception of taxi drivers, people aren't usually in that much of a hurry. Even the chains of fast food shops in Mexico are leisurely by northern standards. When in Mexico, as in Rome, do as the local people do and you'll enjoy it more.

Flying your own plane in Mexico differs only slightly from what you do north of the border. Sure, paperwork can be exasperating and take longer than you expect, but it does get done eventually—and then you're free to sample a myriad of new experiences in "Amigoland."

If you can stand tropical heat and humidity, you'll save a bundle by traveling from mid-May to mid-December when room rates are 20 percent to 45 percent less than during the tourist season.

In subsequent chapters, we'll detail what to take, how to cross the border, and where you might want to go. In general, we'd hope that you fly good equipment that is not in need of repairs. The radios should work, the tires should be relatively new, and the engine should be well within limits. Maintenance is available in Mexico, but parts and overhaul equipment are at a premium. The only bargains we know of on aircraft in Mexico are upholstery, paint, and fuel. (At press time, fuel prices were going up dramatically.)

What follows is intended as an overview of lightplane flying in Mexico. This isn't a tour book that lists every motel and tourist attraction, or for that matter, every airport and every nav/aid, since all these things are subject to change. Procedures, flight techniques, and experiences shared herein are only intended as a background for flying in Mexico. In no way can the authors guarantee that readers will find the same conditions, situations, or prices as those we have described. You fill your own fuel tanks and make your own choices.

We have found Mexico eminently worth visiting with our own aircraft. We hope that you find it the same.

We'd like to acknowledge a great deal of help over many years of flying in Mexico. The Mexican Department of Tourism, and particularly Rolando Torres, veteran of the Los Angeles office, has always been supportive. Bud Lewis, The Lewis Co., Ltd., public relations consultant for the Mexican Department of Tourism, supplied stock photos and background. Ing. Javier Garcia Olave, Aeropuertos y Servicios Auxiliaries, provided information on the ASA airports.

Gil and Bob Gunnell of Gunnell Aviation, Santa Monica, California, have been a constant inspiration with their carefully planned "Fly For Funsters" trips. Francisco Muñoz, pioneer Mexican pilot, spent time with us on the early history of flying in Mexico.

Beldon Butterfield, Editor, and Sergio Perez Rojas of *Guide Magazine* gave permission to reproduce detailed tourist information for several of the larger towns. The Texas State Aeronautics Commission's booklet *Mexican Flight Manual* was a valuable source for material. Gary Wiley of AOPA's Travel Department helped keep us up-to-date on the latest changes in Mexican regulations.

Here's a very deep tip of the sombrero to all of you and to the others that we've mentioned in the body of the book itself. And from a purely personal standpoint, we'd like to thank both Lycoming Engines and Continental Motors for supplying dependable, smooth-running power over some very challenging terrain.

Chapter 1
Preflight

Before you fly yourself to Mexico, try to obtain as much information as you can about the country so you'll have some idea where to go and what to see. The Mexican Department of Tourism has developed a great many fine color brochures on various aspects of travel in Mexico.

Be sure to ask for the booklet *Welcome to the Friendly Skies of Mexico*, prepared by ASA (Aeropuertos y Servicios Auxiliares). (Fig. 1-1). This booklet shows the location of all the ASA government airports and gives an up-to-date listing of the various facilities available when the booklet was prepared. Remember, however, this listing can change just as it does in the U.S. when a nav/aid is changed or an airport temporarily runs out of a certain grade of fuel (Fig. 1-1).

The Tourism Department offices in the United States are listed in the Appendices. A letter or phone call to any of these offices will bring a prompt package of very "rosy" color brochures, which show you the country at its very best. The Mexican consulates (also listed in the Appendices) can supply much of the travel brochure material issued by the government. Stop in and see what they have to offer. When you include "border monetary transactions," tourism is Mexico's biggest foreign exchange producer.

The latest year for which complete statistics are available at this time showed that a total of 3,800,000 foreign tourists visited the interior of Mexico, each spending an average of 10.2 days. Each tourist spent an average of $29.10 per day, for an average of $298.40 during their visit (Fig. 1-2). The U.S. contributes by far the greatest percentage of foreign tourists to Mexico—83.8 percent, with Canada at 4.9 percent second.

Visit your local Auto Club office or write for their surface travel booklets. The AAA *Travel Guide to Mexico and Central America* is revised

Fig. 1-1. Map of Mexico with main airports listed. Many airports now under development are not shown on this map (courtesy Texas Aeronautics Commission).

Fig. 1-2. Every Mexican city has its share of small shops. This tourist shopping area in Ensenada, B.C. is popular with California tourists. Note parking meters.

annually and has the best guide we've found for hotel and restaurant accommodations. On that cab ride in from the airport, it's always comforting to have a specific hotel in mind that's within your budget. Take the booklet along; it isn't heavy.

There are several excellent guide books, including Stephen Birnbaum's 450-page paperback entitled *Mexico* (Houghton Mifflen). See your local bookstore. *Travelers Guide to Mexico* (P.O. Box 6-1007, Mexico 6, D.F. Mexico) at $6.95 is colorful and nightclub oriented; this guide still gives a good overview of what you can expect in the larger towns of Mexico. We picked it up in Mexico City at the tourist book store in the hotel lobby. Don't forget your local library. Take time to browse through their shelves and see what they have to offer.

Depending on your style of travel and your general optimism, you may or may not want to plan a tight itinerary that would include hotel reservations. There is always the chance of a weather or operational delay that could throw your schedule off or present an undue urgency to "get there" rather than playing it safe.

There are many hotels in Mexico and several chains, including the Holiday Inn, that do a fine job. We had excellent, reasonably priced accommodations at La Reforma Hotel in Mexico City, which is a member of a Mexican chain. Unique in Mexico is the government-sponsored chain of El Presidente Hotels. We've stayed at several of them with very satisfactory results. El Presidente Hotels have the added charm of being strictly first-class Mexican from top to bottom. For advance reservations, there are two U.S. toll free numbers: California—(800) 542-6028; nationwide—(800) 854-2026. The present list of El Presidente Hotels includes:

- Baja California
  - La Pinta Ensenada
  - San Quintin
  - Catavina
  - El Presidente Sur, La Paz
  - Loreto
  - San Jose del Cabo (open 1982)
  - Guerrero Negro
  - San Ignacio
  - La Pinta Loreto
- Northeast
  - Chihuahua
  - Durango
  - Torreon
- Mexican Caribbean
  - Cancún
  - Cozumel

Central Region
- Ixtapa-Zihuatanejo
- Chapultepec, Mexico City
- Del Prado, Mexico City
- Zona Rosa, Mexico City

Southeast
- Oaxaca
- Campeche
- Villahermosa
- Chetumal

AVIATION INFORMATION SOURCES

In the nuts-and-bolts of aviation, write the Texas Aeronautics Commission (P.O. Box 12607, Capitol Station, Austin, Texas 78711) for their free *Mexican Flight Manual*. This excellent publication is updated periodically and is 130 pages of good reading—and the price is right!

The most detailed of the flying guide books is Arnold Senterfitt's *Airports of Mexico & Central America*. Senterfitt has been dedicated to reporting on Latin American aviation for more than 20 years. His 560-page book ($24.95) is painstakingly detailed and as accurate as they come. It's available through your local FBO or from Sky Buys, Box 4111, San Clemente, CA 92672.

AOPA's Flight Planning Department (Aircraft Owners & Pilots Association), 421 Aviation Way, Frederick, MD 21701 has a package of detailed Mexican tips and flight planning available for its more than 260,000 members. AOPA is in a position to know any last-minute regulation changes, frequently before they are officially announced.

Two more publications you might wish to send for are *Custom Hints for Returning U.S. Residents* (Bureau of Customs, Washington, D.C. 20226) and *Answers to Your Travel Questions About Mexico* (Mexican Government Tourism Department, Reforma and Lafragua, Mexico City, and other Mexican Government Tourism offices. Both publications are free of charge.

FEE FOR FLIGHT PLANS

A Mexican directive dated February 17, 1981, in the Mexican Congressional Journal established a fee of 150 pesos for every Mexican VFR flight plan and 400 pesos for each IFR flight plan. At the 1981 peso/dollar exchange, that's $6.52 and $17.40 U.S. currency, respectively. (Be sure to check the current exchange rate.) The original announcement detailed that this new and completely unexpected fee would initially apply at Tijuana, Monterrey, Monterrey del Norte, Mazatlán, Acapulco, Merida, and Guadalajara. The Mexican announcement also stated that this new fee can be expected at all airports where flight plans are issued at a later date.

This new flight plan fee structure is inconsistent with the efforts of the Mexican Tourist Bureau to encourage visitors to Mexico. As soon as the initial announcement was made, members of tourist services and hotel chains began making every effort to have this added fee either cancelled or incorporated with fuel costs as have been other fees for services. Readers are urged to contact their nearest Mexican Tourist Bureau or Mexican Consul to ascertain the status of the flight plan charge before departing for a south-of-the-border trip. Obviously, it will be extremely unpopular with general aviation pilots planning to visit Mexico.

When you fly yourself to Mexico, take cash and travelers' checks. Unless you have so verified in advance, don't even count on the major credit cards being accepted for food and lodging. On our most recent visit to Baja, California, we found that a majority of the hotels on the tip at San Jose del Cabo and Cabo San Lucas were *not* accepting credit cards. Whether this trend is general or only in southern Baja, California, has not been ascertained. However, there is nothing that will louse up your trip quicker than having to figure the cost of each move you make as it applies to the cash (or lack of) you have in your pocket when you had depended on credit card usage.

We've said it elsewhere, but it bears repeating. *Do not count on paying for your av/gas with credit cards*! On a visit to La Paz (in 1981), no charge slips were available on a busy Saturday, so we paid cash (100 octane, $1.54 per gallon). Yet two days later, we were able to use our Visa card for fuel at the same pumps. Other pilots report that Mexicali is fairly consistently "out of charge slips."

CHARTS

If you're like we are, you get your charts well in advance and study them at your leisure. Aviation has been improving in Mexico so rapidly recently that only the latest charts are really useful. Your list of aeronautical charts should include:

Sectional Charts

- ☐ Phoenix
- ☐ El Paso
- ☐ San Antonio
- ☐ Brownsville

World Aeronautical Charts (NOS)

- ☐ WAC CH-23

Operational Navigation Charts (DMAAC)

- ☐ ONC H-22
- ☐ ONC H-23
- ☐ ONC J-24
- ☐ ONC K-25

DOD Flight Information Publication (DMAAC)

- ☐ Enroute Low Altitude L-1 and L2, Caribbean and South America
- ☐ Enroute High Altitude H-1, Caribbean and Northeast Coast of South America

ICAO STANDARD

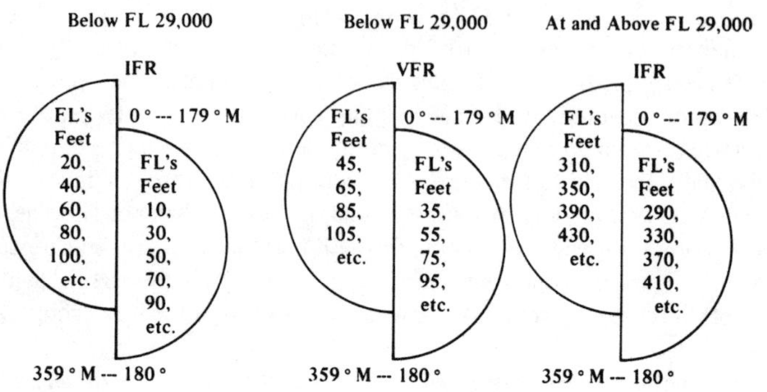

Exceptions:

1. VFR flights are not authorized above FL 200.

2. As an exception to ICAO, Southbound flights between ALBUQUERQUE and MEXICO and flights between HOUSTON and MEXICO at FL 290 or above adhere to FLs 310, 350, 390, etc. For flights between MEXICO and ALBUQUERQUE or HOUSTON at FL 290 or above adhere to FLs 290, 330, 370, 410, etc.

Altimeter settings QNH below 140 - QNH indicates cruising altitude.

Altimeter settings QNE above 150 - QNE indicates cruising level.

Fig. 1-3. Pictorial display of off-airway altitudes in Mexico.

Colored Airway Chart of Mexico

In addition to the above, JNC (Jet Navigation Chart) #46 on a scale of 1:2,000,000 covers most of Mexico and is suited for planning or as a wall map in the office or den. Available from DMAAC, this chart does not cover most of Yucatán and also misses the Guatamala border area at Tapachula (Fig. 1-3).

Pilots with ADF (radio compass) equipment will find *Oliver's ADF Directory* useful. It can be ordered from G & O Publishing Co., P.O. Box 225, Garwood, Texas 77442. We carry one in our plane at all times. In addition, you'll find a partial list of Mexican Commercial Broadcast Stations in the Appendices of this book.

While the following publications are not essential for VFR flying in Mexico, you may want to add some or all of them to your flight case:

DOD Flight Information Publication

Enroute Low and High Altitude radio facility and airway charts, Caribbean and South America.

Distribution Division, C-44
National Ocean Survey
Riverdale, Maryland 20840

One time issue charts and supplement $2.55
Annual subscription $15.20

DOD Flight Information Publication Supplement (for use with Caribbean and South American FLIPS)

A directory of aerodromes with 3000 ft. or longer runways and others, radio aids to navigation, data for communications, DF, ILS, and radar available a city/aerodrome cross reference list, VFR aerodrome sketches, area IFR, VFR, emergency and position reporting procedures, special notices, cruising altitude diagrams, and conversion tables.

Available from National Ocean Survey, as above. Brief justification is necessary to substantiate requirements.

International Flight Information Manual

Foreign entry requirements, a directory of aerodromes of entry, including operational data, and pertinent regulations and restrictions, passport, visa and health requirements for each country.

Superintendent of Documents
U.S. Government Printing Office
Washington, D.C. 20402

Annual subscription price:
$6.00 in the U.S.

International NOTAMS

Supplements International Flight Information Manual

Superintendent of Documents
U.S. Government Printing Office
Washington, D.C. 20402

Annual Subscription: $18.00 in the U.S.

Manual de Facilidades y Procedimientos de Navegacion Aerea

Informes Aeronauticos "Bernal"
Donato Guerra 1-609
Mexico, 1, D.F.

Contains (in English) NOTAMS, aerodromes, radio navigation aids and control tower hours of operation, communications and navigation aid frequencies, airport directory and sketches, instrument approach plates, list of commercial broadcast stations, and aviation weather reporting stations in Mexico.	Basic manual: $12.00 Monthly revisions and weekly bulletin service: $24.00 a year.
Mexico's Airmans Information Manual (English & Spanish)	At Mexican airports of entry

BORDER-CROSSING AIRPORTS

You'll want to study Mexico's new regulations (issued December 15, 1979) that apply to visiting aircraft. The new regulations require that all one-and two-engine piston aircraft, excluding jet or turbine aircraft, enter and depart from Mexico at one of the following airports of entry:

Northern Zone

For flights from or to the northern border of Mexico: Tijuana/General Abelardo L. Rodriquez, B.D.; Mexicali/General Rodolfo S. Taboada, B.C.; Nogales, Son.; CD. Juarez, Chih.; CD. Acuna, Coah.; Piedras Negras, Coah.; Nuevo Laredo, Tamps.; Reynosa, Tamps.; and Matamoros, Tamps. Subsequently, Hermosillos La Manga International was added to the list (Fig. 1-4).

South and Southeastern Zones

For flights from or to the south and southeastern borders of Mexico: Tapachula, Chis.; Chetumal, Q.R.; Cozumel, Q.R.; and Cancún, Q.R.

The crew and passengers who travel on these aircraft will comply with the requirements and formalities of Customs, Immigration and Health at the international airports through which they will enter and depart the Republic of Mexico.

This regulation is applicable only to private aircraft of foreign nationality and foreign registration of any type whose maximum capacity is 16 seats for passengers and for which there is no transportation charge. These aircraft, can only transport people who have been invited and under no circumstances may carry any cargo, merchandise, or articles of any kind that are not for personal use of the crew or their guests.

The 1979 basic change in Mexican flights requiring an initial landing at the nearest available approved Mexican airport of entry does not apply to

Fig. 1-4. Control tower and twin Comanche at Hermosillo, Sonora. This airport was added to the list of approved airports of entry for general aviation aircraft.

jet, turbine, or air carrier aircraft or other aircraft having a seating capacity of more than 16 people.

When the capacity of the aircraft is more than 16 seats, application for approval must be applied for at least five working days before the date of the scheduled trip and written permission must be obtained from Directorate of Civil Aeronautics. Jet or turbine aircraft can and should enter and depart from Mexico at any of the aerodromes listed below, except Monterrey/Del Norte:

Acapulco, Gro.; Cancún, Q.R.; CD. Acuna, Coah.; CD Juarez, Chih.; Cozumel, Q.R.; Chetumal, Q.R.; Chihuhua, Chih.; Guadalajara, Jal; Guaymas, Son.; Hermosillo, Son.; La Paz, B.C.; Manzanillo, Col.; Matamoros, Tamps.; Mazatlán, Sin.; Mexicali, B.C.; Mexico, D.F.; Merida, Yuc.; Monterrey, N.L.; Nogales, Son.; Nuevo Laredo, Tamps.; Puerto Vallarta, Jal.; San Jose, Del Cabo, B.C.; Tampico, Tamps.; Tapachula, Chis.; Tijuana, B.C.; Torreon, Coah.; Veracruz, Ver.; Zihuatanejo, Gro.; and Piedras Negras, Coah.

HOMEBUILTS

An ever-increasing number of homebuilts (they're called "housebuilts" in Mexico) are now flying in the U.S. and crossing into Mexico. Pilots use their U.S. "N" registration number and an identifying name for their type of aircraft and seem to proceed with all the same treatment provided production aircraft. For example, a sign on the wall of one of the

Mexican border flight offices identifies a VariEze as the "Evans VariEze." Evans was the first builder of this particular design to cross at this border crossing (Fig. 1-5).

Norm Spitzer, another VariEze builder from Berkeley, California, took his homebuilt to Guadalajara and returned with no problem. Spitzer's was the fifth VariEze to clear in through Mexicali, and he reported "one of the nicest trips I've ever had." He carried MacAfee & Edwards, Mexican Insurance. Because of the unusual canard design with the small wing in front, Spitzer reported that one gas man in Mexico thought it was a flying bomb and didn't want to fuel it. There were so many curious visitors around his new little homebuilt that Spitzer could hardly get to it at most of the airports.

INSURANCE

The operator of an aircraft must have insurance that will guarantee any damages caused to third parties on the ground due to operations effected in Mexican territory, according following value:

Aircraft up to 5,000 kgs gross weight, the amount of	$60,000.00
Aircraft up to 20,000 kgs. gross weight, the amount of	$150,000.00
Aircraft up to 40,000 kgs. gross weight, the amount of	$600,000.00

This insurance should be arranged with a Mexican insurance company prior to arrival in Mexico. While this required liability policy is supposed to

Fig. 1-5. Four VariEze homebuilts are shown parked on the ramp in front of the Tijuana, B.C. control tower during a group flight into Mexico. This group of VariEze builders make frequent flights into Mexico. Homebuilts are handled just like any other aircraft with U.S. registration numbers (courtesy Al Coha).

be available on the airport of entry, we have yet to find it except at the end of a long taxicab ride into a border town. Then you must find an approved insurance vendor who is open and purchase the policy. It is easy to obtain in advance and will help start your trip properly.

While there are many to choose from, here are two insurance companies that we have used personally. Raul Martinez, Mexair Insurance Brokers, 12601 Venice Boulevard, Los Angeles, CA 90066; (213) 398-5797, can supply Tourist Cards, fishing licenses, hunting licenses and gun permits, as well as a health certificate for your pet, in addition to aircraft liability insurance. In 1980, Mexair's aircraft premiums, based on $30,000/$60,000 U.S. currency for Bodily Injury Liability (excluding passengers and crew) and $25,000 U.S. currency for property damage, were based on 5 percent of the total annual premium for the first day. For example, $6.88 plus $1.00 to write the policy for one day with 6 percent California tax added, or a total of $8.35. Ten days coverage would be $15.64.

One of the oldest companies handling Mexican aircraft liability insurance is MacAfee & Edwards, 3105 Wilshire Boulevard, Los Angeles, CA 90012; (213) 388-9674. MacAfee & Edwards issues a plastic wallet-sized "Mexicard" for a one-time fee of $5.00. More than 200,000 "Mexicards" have been issued. Even though the "Mexicard" is mailed out the same day the application is received at MacAfee & Edwards, mail being what it is, it usually takes about two weeks. After receipt of this "insurance credit card," the requisite Mexican insurance for your aircraft can be activated by a phone call. In all the trips we've made to Mexico using this card, we have never had a problem. The main advantage of this "plastic insurance" is that you can decide to go to Mexico on the spur of the moment and obtain your insurance with a phone call. Recently, when AOPA Vice President Bob Warner and I finished a meeting in San Diego and looked for a place for lunch, I suggested Ensenada in Baja California since Warner had never before flown into Mexico. After a quick phone call to MacAfee & Edwards, we were on our way with the required insurance.

Any expensive foreign-made equipment such as cameras or binoculars should be registered with U.S. Customs before departing from the U.S. We carry a copy of our camera insurance policy that lists the serial numbers of all cameras and lenses. This documentation has been adequate to date. Write or phone your nearest Customs Office for a copy of the pamphlet *Know Before You Go*, (U.S. Government Printing Office #726-216). Also an excellent *U.S. Customs Guide For Private Flyers* can be obtained from the Customs Office or by writing The Treasury, U.S. Customs Service, Washington, DC 20229.

If your trip to Mexico includes a motorbike or two carried in your airplane, be sure to list the bikes, complete with all pertinent numbers, when you apply for your aircraft insurance. Also list the number of days you plan to use to use the bikes in Mexico so that this may be listed on a separate insurance policy.

If you plan to take a cat, dog, or other pet, be sure to obtain a Mexican vaccination certificate, in duplicate, and have it filled out before arriving at the Border. Pet certificates are available from the Mexican Consul or from most insurance companies; there is a small fee.

WHAT TO CARRY IN THE PLANE

Make certain that all the paperwork on your airplane is up-to-date and where you can find it. Mexican manifests, unlike those in the U.S., include the serial number of the aircraft and border officials want to see that aircraft registration form. If you are flying a rented aircraft or one that belongs to a friend or a partnership, make sure that you have written authorization, preferably on letterhead and notarized, that you (the pilot) are authorized to fly this particular aircraft in Mexico. If this is a one-time trip, a list of dates might well be included.

We carry a couple of quarts of the type of oil our engine has in its crankcase, since the selection in Mexico is limited. Take a chamois and plastic funnel if you plan refueling at any but the largest towns (Fig. 1-6). Carry your own tiedowns, both ropes and some sort of anchor. Empty canvas sacks are adequate, as are three-foot aluminum extrusions (our favorite) with three stout nylon ropes. If your bird doesn't have an internal rudder lock, we'd suggest taking one.

We suggest you carry a minimal repair kit with pliers, two types of screwdrivers, a crescent wrench, and one or two spare spark plugs. It's just good business. Tires and/or tubes are a personal option. If you're headed for paved runways and your tires aren't already on threads, then you may be safe without them. On smaller gravel or dirt strips, a spare could be useful.

A space blanket or other foil sheet will protect your instrument panel from the heat while your ship is parked and will keep inquisitive eyes from seeing your cockpit. It will also serve as a lightweight thermal blanket with protection against hot or cold in case of a night away from town. The two space blankets we carry date back to about 1969, and they haven't worn out yet.

Take a clean cloth or two and windshield cleaner since few refueling facilities provide this service. And we *strongly* suggest that during refueling you stay with your aircraft to make certain you are getting the correct grade of fuel. Check the amount with your own personal estimate, just as you would stateside.

Put together some sort of a survival package, just in case you have a problem. We'd recommend including one gallon of water (more if your route is to be over the arid midsection of Mexico where roads and civilization are sparse); signaling mirror; axe and/or strong knife; first aid kit; waterproof container of matches; sunburn lotion and sunglasses. Beyond that, carry the same thing you would on a long stateside flight.

We carry snacks for "in-flight energy;" choose your own favorites. A supply of peppermint sticks or other rock candy can be most useful in cementing relations with the youngsters that swarm around the smaller

Fig. 1-6. Aircraft line up for fuel at Serenidad, B.C. Fuel here is filtered as it comes out of large gravity tanks so that a chamois skin has not been necessary.

Fig. 1-7. "Stop. Precaution Landing Field for airplanes, Rancho Buena Vista." Small resort airports like this one have little control. You can expect to find youngsters, dogs, cattle, or whatever in the area when you land.

airports. All ASA fields are well controlled and you won't have this contact with a crowd of youngsters descending on your ship as you shut down (Fig. 1-7).

If your flight plan calls for only a quick fuel and bladder stop, you may want to pack a lunch or have your hotel put together a box lunch. In-flight liquids are fine but should be consumed with some discretion since a mismatch of long-range fuel tanks and short-range passengers can really foul up your preflight planning.

PACKING

Since climates in Mexico range from cool to humid tropical, you can pack just about anything you wish. Comfortable walking shoes for flying and airport use are a must. Make sure that they are broken in and that you have soft socks that don't bind at toe, heel, or ankle. A light jacket and wool sweater will handle most chilly conditions and eliminate packing a heavy coat in your baggage.

If you're headed for Puerto Vallartz, Veracruz, or farther south, take along whatever personal anti-perspiration treatment works for you—spray, solid, or roll-on.

Boots may be useful if you're going off the beaten path. Insects can

Fig. 1-8. Eat and drink at places that look clean. The outdoor bar at the famed Hotel Cabo San Lucas has hand-carved, contoured bar stools.

easily conquer low-heeled shoes and leave a lasting impression. However, for most tourist resorts, such ankle protection is not necessary.

Include an English/Spanish dictionary, even a small one, unless you are proficient in the language. A simple pocket calculator is the best solution to the dollar/peso, gallon/liter, pound/kilogram hassle.

We suggest you ask your local physician for what to take in your traveling kit to combat "Montezuma's revenge." While some doctors are not recommending a "binding" prescription, we still carry a few Lomatil tablets, just in case. We do, however, heed the following suggestions from experts: Don't pre-medicate; eat at places that *look* clean (Fig. 1-8), water is purified in the better restaurants and hotels. When in doubt, ask for bottled water. Don't eat unpeeled fruit or vegetables; for mild diarrhea, try a bland diet, skip the booze for a couple of days, and think about using Kaopectate, Pepto-Bismol, etc.

From the medicine chest, keep personal prescriptions in their original containers with the labels intact so that any inspection is simplified. Do not mix vitamin pills and/or whatever, since it is hard to prove what you are actually taking. We have never had a problem with either prescription or over-the-counter drugs, but any white vitamin C tablet looks just about like any other white tablet and border inspectors are careful.

If you're *really* compulsive and have made up your mind to fly yourself to Mexico right now, we'd suggest reading the final chapter in this book on returning to the U.S. before heading for the airport.

TOURIST PILOT'S CHECKLIST

1. Proof of Citizenship:

a. U.S. citizen—birth certificate, voter's registration.

b. Naturalized U.S. citizen—naturalization papers or U.S. passport.

c. Alien resident—valid passport and alien registration card.

d. Other—valid passport.

2. Valid International Certificate of Inoculation and Vaccination, if required.

3. Tourist Card.

4. General Declaration.

5. Flight Plan.

6. Aircraft documents:

a. Proof of ownership—title or notarized affidavit from owner authorizing trip to Mexico.

b. Aircraft Registration Certificate.

c. Aircraft Airworthiness Certificate.

d. Aircraft Weight and Balance Data.

e. Aircraft Radio Station License.

7. Pilot Documents:

a. Pilot's Certificate.

b. Medical Certificate.

c. Radio License.

8. Special Mexican Aircraft insurance.

9. Aeronautical Charts.

10. Adequate survival gear and supplies.

11. Tiedown kit and chamois.

12. Medicines—see your family physician.

13. Declare expensive foreign made cameras and other items before entering Mexico.

14. Adequate supply of Mexican currency for tipping and making change.

Chapter 2
Mexican Border Crossing

The first time you call a Mexican tower on your radio and a slightly accented voice clears you to land, you know that you're in for an adventure of one kind or another. No matter how many times you make that border crossing, there is still that sense of breaking loose and something different. In years past a lightplane flight to Mexico was certain to be an adventure—one fraught with bribes and all sorts of official roadblocks. That hassle fortunately is long gone for the most part. Today's fly-in visitor to Mexico can expect to be treated courteously, pay no more than predictable fees, and have a delightful trip (Figs. 2-1, 2-2).

This is not to say that you can't meet the wrong border official on the wrong day at the wrong airport and have a hard time. But for the most part, today's procedures are good and getting better.

Since times and regulations change, we strongly recommend that your first visit to Mexico (or even your first visit across the border at an entry point new to you) be preceded by a landing on the corresponding U.S. airport of entry for a nose-to-nose update on today's requirements, what airports may be out of fuel, and what specific problems have developed in that particular area.

One of our favorite pre-border crossing spots is the FAA FSS (Flight Service Station) at Imperial, California, on the way to Mexicali. This FSS keeps a status board on any changes in procedures as well as specific Mexican airports and resort data received from debriefed returning pilots. In addition, the airport at Imperial has a good "airport motel," restaurant, and bar right on the field for those pilots who want to get off early and touch down at Mexicali at 0700 when their offices open. The Imperial FSS uses the pictured put-in-your-pocket briefing card (Fig. 2-3).

Fig. 2-1. Control tower at the Mexican border airport of Tijuana, B.C. Military aircraft parked on the ramp is a Swiss Pilatus PC-7 turboprop two-seater. All the services needed by a general aviation pilot in Tijuana are on the first floor of the tower building.

FAA FLIGHT SERVICE STATION ADVISORY

The Imperial FSS provides the accompanying letter to airmen:

Mexican Border Crossing Requirements and Information

"Flying to Mexico is not as simple as walking or driving across the border, but if you plan ahead and visit your nearest Flight Service Station, a trip by air can be very enjoyable. Border Flight Service Stations are prepared to assist airmen in planning and preparing for flights to and from Mexico. The following information has been prepared to assist in your flight planning:

"Federal Aviation Regulation (FAR) 91.84 requires all aircraft crossing the U.S./Mexican border to be on a U.S. Flight Plan. A border crossing flight plan can be filed with any Flight Service Station. If you wish to file a flight plan by radio with Imperial FSS, do so at least 30 minutes prior to crossing the border. A request for Customs and Immigration notification may be made at this time. Pilots desiring to file a round-robin flight plan with Imperial FSS for Search and Rescue purposes may combine this information with their border crossing flight plan if returning through Calexico, California International Airport. All round-robins must provide the name and telephone number of a person in the United States with first-hand knowledge of your itinerary who will be available to aid in a communication search if necessary." *(Note: The authors make it a habit to close their U.S. Border Crossing flight plan by radio before landing at the Mexican Airport of Entry. This flight plan will also be closed in writing on the ground as the first order of business, but this message may or may not be transmitted back to the U.S. side.)*

"All aircraft must use an Airport of Entry to enter or leave Mexican Territory. After the first landing in Mexico, permission must be obtained

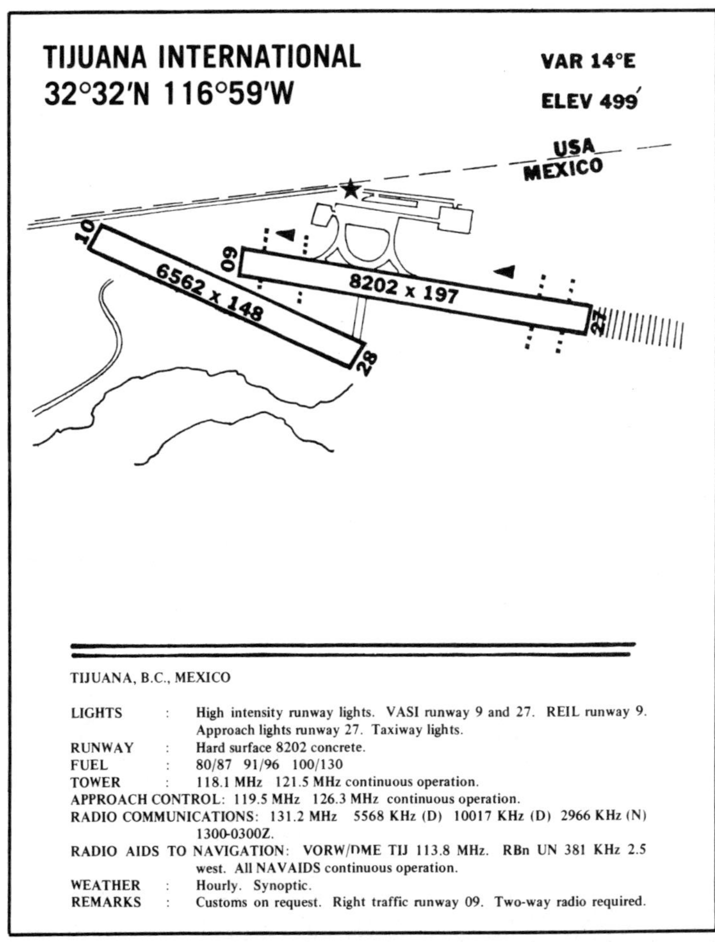

Fig. 2-2. Layout of the Tijuana International Airport. Runway 10/28 is not used for international travel. All frequencies listed may be obsolete. Not for navigation.

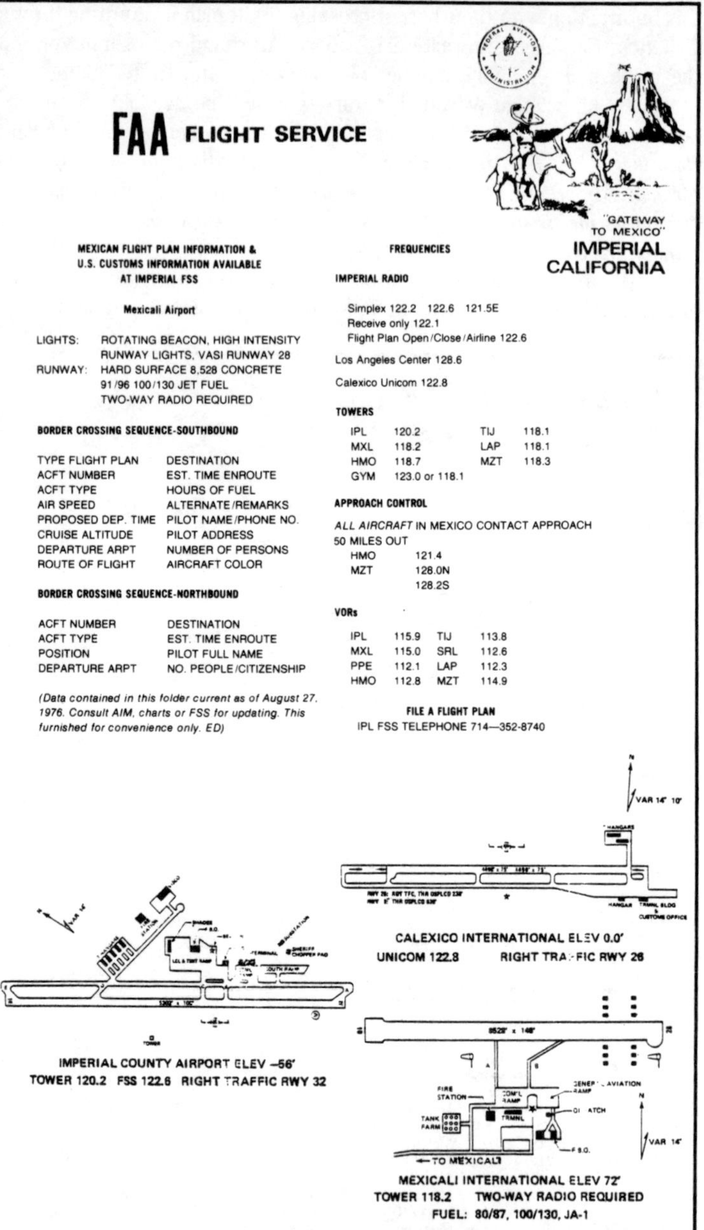

FAA FLIGHT SERVICE

"GATEWAY TO MEXICO"

IMPERIAL CALIFORNIA

MEXICAN FLIGHT PLAN INFORMATION &
U.S. CUSTOMS INFORMATION AVAILABLE
AT IMPERIAL FSS

Mexicali Airport

LIGHTS:	ROTATING BEACON, HIGH INTENSITY RUNWAY LIGHTS, VASI RUNWAY 28
RUNWAY:	HARD SURFACE 8,528 CONCRETE
	91/96 100/130 JET FUEL
	TWO-WAY RADIO REQUIRED

BORDER CROSSING SEQUENCE-SOUTHBOUND

TYPE FLIGHT PLAN	DESTINATION
ACFT NUMBER	EST. TIME ENROUTE
ACFT TYPE	HOURS OF FUEL
AIR SPEED	ALTERNATE/REMARKS
PROPOSED DEP. TIME	PILOT NAME/PHONE NO.
CRUISE ALTITUDE	PILOT ADDRESS
DEPARTURE ARPT	NUMBER OF PERSONS
ROUTE OF FLIGHT	AIRCRAFT COLOR

BORDER CROSSING SEQUENCE-NORTHBOUND

ACFT NUMBER	DESTINATION
ACFT TYPE	EST. TIME ENROUTE
POSITION	PILOT FULL NAME
DEPARTURE ARPT	NO. PEOPLE/CITIZENSHIP

(Data contained in this folder current as of August 27, 1976. Consult AIM, charts or FSS for updating. This furnished for convenience only. ED)

FREQUENCIES

IMPERIAL RADIO

Simplex 122.2 122.6 121.5E
Receive only 122.1
Flight Plan Open/Close/Airline 122.6

Los Angeles Center 128.6

Calexico Unicom 122.8

TOWERS

IPL	120.2	TIJ	118.1
MXL	118.2	LAP	118.1
HMO	118.7	MZT	118.3
GYM	123.0 or 118.1		

APPROACH CONTROL

ALL AIRCRAFT IN MEXICO CONTACT APPROACH 50 MILES OUT

HMO	121.4
MZT	128.0N
	128.2S

VORs

IPL	115.9	TIJ	113.8
MXL	115.0	SRL	112.6
PPE	112.1	LAP	112.3
HMO	112.8	MZT	114.9

FILE A FLIGHT PLAN
IPL FSS TELEPHONE 714—352-8740

Fig. 2-3. Flight Service Station information sheet prepared by the Imperial, California FAA station. This is typical of the type of information issued by FAA facilities along the border. However, radio frequencies were valid only when this card was prepared in November, 1978.

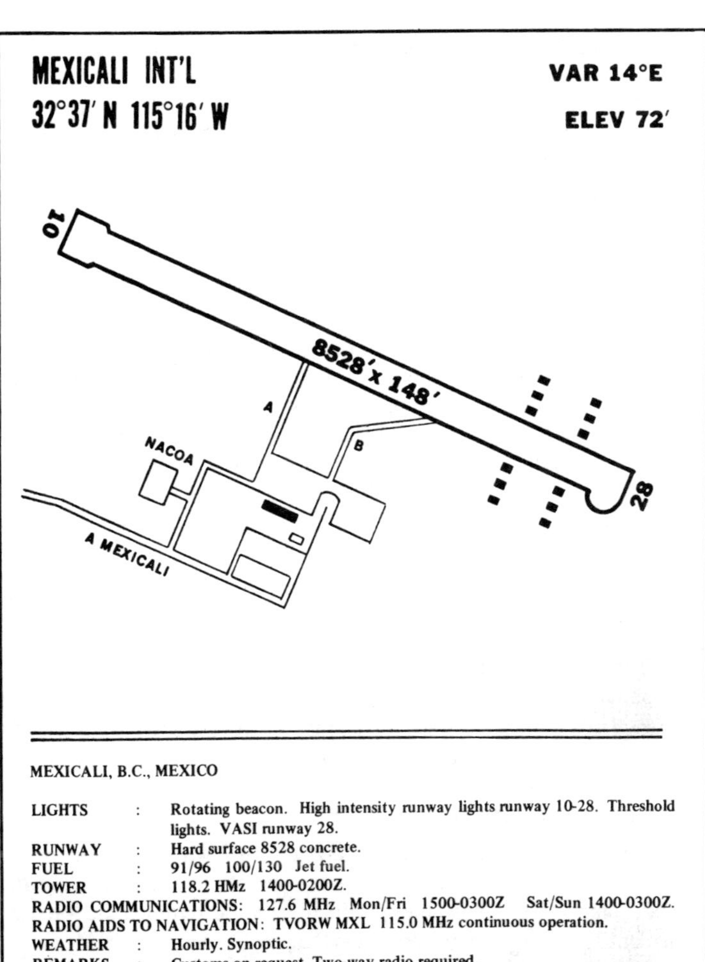

Fig. 2-4. Typical airport layout at a Mexican Airport of Entry. This is the size of the Mexicali International Airport. All services and radio frequencies are subject to change. Not to be used for navigation.

from the Comandante of the airport to pick up additional or substitute passengers or crew members for flights to other cities within Mexico or for flights leaving Mexico. At some locations, permission may not be granted." *(A list of the Airports of Entry in Mexico is carried in the previous chapter.)* (Fig. 2-4.)

Clearing Mexican Customs and Immigration

"Upon arrival the pilot will be directed to a designated parking area. The pilot and passengers must then proceed to the Airport Operations

Office. At this time an arrival report will be filed and a new flight plan filed for the next leg(s) of flight. A written flight plan is required for all flights within Mexico. IFR flight plans are required for all night flights.

Flight plans must be closed in writing as soon as possible after arrival at locations that provide air traffic service. This may be the airport of entry when departing Mexico if all flight operations in Mexico are from airports without air traffic service."

Documents Required

"Citizenship: Each person must have a passport, birth certificate, (or a certified copy), military discharge papers, voter's registration card, or a notarized statement swearing that you are a United States Citizen." *(The authors suggest that you be certain to carry sufficient proof of U.S. citizenship to permit entry back into the U.S. on your return.)*

"Minors (under 18): In addition to proof of citizenship, minors must have a notarized letter from one or both parents, authorizing them to make the trip. If one parent is present, a notarized letter from the other parent is required.

"Pilot: Proof of citizenship, current airman and medical certificates, personal radio license, logbook and proof of Mexican insurance for the aircraft.

"Aircraft: Documents must include aircraft radio license, aircraft registration, airworthiness certificate, and operator's manual. If the aircraft is not owned by the pilot, it is suggested that written authorization for flight into Mexico from the owner, club, etc., be included with these documents."

AT THE AIRPORT OF ENTRY

Normally, the man at the initial operations desk will ask for your pilot's license to copy down proper spelling of your name, the number of the license, home address, and other pertinent data. Be sure that the Mexican flight plan you receive has your name in the proper sequence. In Mexico, it is proper to use the middle name (which in Mexico is the father's surname) as an integral part of your official name. The last or "third name" in the Mexican name is usually the mother's maiden name, or so it has been explained to us. Thus the name Donald Chase Downie can come out on your Mexican flight plan and subsequent G.H.C. 001 ("General Declarations" form) as pilot's name, "Chase." This has happened more than once (Fig. 2-5). That'll be small help if you show up missing and your family instigates a search for a pilot named "Downie." Remedy: Return the form if it is filled out incorrectly and ask the Mexican official to rewrite it with your last name last. Normally, the error will be "x'd" out on the typewriter, corrected, and initialed in no more than a minute.

The Mexican "General Declaration" form is numbered G.H.C. 001. This form must be kept on board the aircraft and shown to Mexican authorities upon request. The aircraft must leave the country carrying the same crew (and passengers) listed when the form was filled out. Any change in pilots and/or passengers while in Mexico should be avoided if at all possible. If any such change is necessary, a signed statement of the change and the reason for it must be obtained from the Airport Comandante where the change took place. For example, if one of your passengers is taken ill in Mazatlán and wishes to return to the U.S. on a jetliner, a signed statement of what happened should be obtained by the pilot in command of the private airplane from the Airport Comandante at Mazatlán to show at the border on return or at any other enroute stops. It isn't like the U.S., where you can change crew and passengers with no authorization. However, the U.S. FAA requires similar documentation from Mexican pilots flying Mexican aircraft in the U.S. (Figs. 2-6, 2-7).

In the course of researching this book and making Mexican border crossings for many years, we have frequently asked Mexican airport comandantes why there was no standardization for incoming U.S. pilots so that they could stop at "Desk No. 1" and then follow a red, green, or blue line through the subsequent steps required in clearing into Mexico. The reply to our query has been that not all border airports are set up in the same way at the same time of day. During slack periods or at mid-week when traffic may be slack, the Customs man may do the job of the Immigration man or vice versa. In any event, the best you can do is to follow directions given at the time of your arrival (Fig. 2-8).

The pilot and his/her passengers will go to the Customs and Immigation Office where Tourist Cards are stamped and one copy returned. For

S.C.T.

PLAN DE VUELO

LUGAR: MZT FECHA 26-5-80 VFR IFR

SECCION SUPERIOR: PARA IFR LLENE TODOS LOS DATOS. PARA VFR TODOS, EXCEPTO LOS SOMBREADOS

A) IDENTIFICACION Y/O NUMERO DE VUELO	B) TIPO DE AERONAVE	C) MATRICULA	D) NIVELES DE CRUCERO Y RUTA
PRIV	C-120	N2672D	55'

E) TIEMPO ESTIMADO ENTRE ESCALAS	HORA DE SALIDA F) PROPUESTA	G) EFECTIVA
1+50		

H) VELOCIDAD VERDADERA	I) AEROPUERTO ALTERNO (S)	J) RADIO FRECUENCIAS A BORDO TRANS	RECEPC	K) COMBUSTIBLE A BORDO HORAS	MINUTOS
				5	00

L) NOMBRE DEL COMANDANTE	M) NUMERO LICENCIA	N) DOMICILIO	O) COLOR DE LA AERONAVE	P) DESTINO FINAL
DONALD CHASE D.	58160	U.S.A.	AVE BCO. AZUL	PVR

OBSERVACIONES: GHC - AC A # 2962
3 PAX

COMANDANTE DE LA AERONAVE DESPACHADOR COMANDANCIA DEL AEROPUERTO

ORIGINAL PARA EL PILOTO: COPIA PARA EL DESPACHADOR: COPIA PARA LA COMANDANCIA

Comandancia Aeropuerto Federal MAZATLAN, SIN.

Fig. 2-5. This copy of a VFR flight plan from Mazatlán to Puerto Vallarta lists the pilot's name as Donald Chase D. The Mexican General Declaration Form G.H.C. 001, which must be turned in before departing from Mexico, initially had this same mistake which was copied by Mexican dispatchers for the remainder of the trip. The "3 PAX" under "Observaciones" refers to three passengers.

Fig. 2-6. A few Mexican airports like Ensenada, B.C. are "joint use" fields where the Mexican Air Force handles dispatching. Here a Mexican Air Force pilot serves as dispatcher. The posted tower frequency of 119.75 MHz was not yet in operation and the U.S. Unicom frequency of 122.8 was used at the time of this visit.

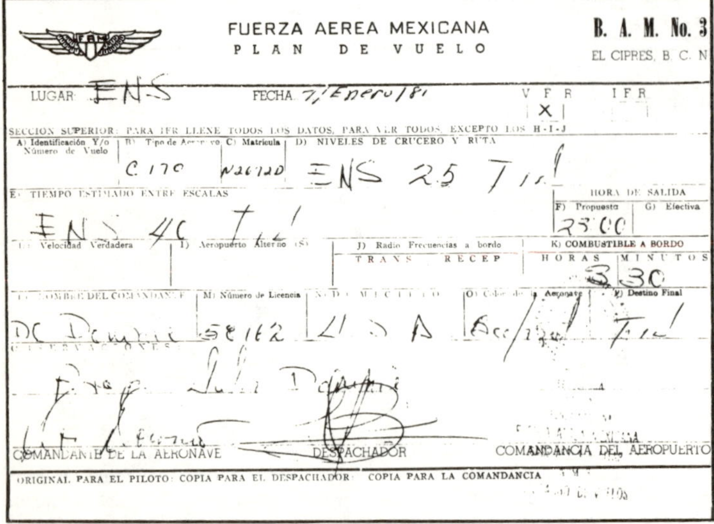

FUERZA AEREA MEXICANA
PLAN DE VUELO

B. A. M. No. 3
EL CIPRES, B. C. N

LUGAR ENS FECHA 7/Enero/81 VFR X IFR

SECCION SUPERIOR: PARA IFR LLENE TODOS LOS DATOS, PARA VFR TODOS, EXCEPTO LOS H-I-J

A) Identificación Y/o Número de Vuelo | B) Tipo de Aeronave | C) Matrícula | D) NIVELES DE CRUCERO Y RUTA

C 170 | N2672D | ENS 25 Tij

E) TIEMPO ESTIMADO ENTRE ESCALAS: ENS 40 Tij

HORA DE SALIDA: F) Propuesta 2300 | G) Efectiva

H) Velocidad Verdadera | I) Aeropuerto Alterno (S) | J) Radio Frecuencias a bordo: TRANS RECEP | K) COMBUSTIBLE A BORDO: HORAS 3 MINUTOS 30

L) NOMBRE DEL COMANDANTE | M) Número de Licencia | N) | O) Color de la Aeronave | P) Destino Final

OBSERVACIONES

COMANDANTE DE LA AERONAVE | DESPACHADOR | COMANDANCIA DEL AEROPUERTO

ORIGINAL PARA EL PILOTO | COPIA PARA EL DESPACHADOR | COPIA PARA LA COMANDANCIA

Fig. 2-7. Mexican Air Force flight plan issued for a flight between Ensenada and Tijuana in Baja California. Note that on this flight plan, the pilot's proper last name is listed.

visitors arriving without tourist cards, these can be obtained at this office. Here you will need the citizenship documents mentioned earlier in this chapter.

There are two different kinds of tourist cards for Mexico. Both are good for 180 days and both are issued free of charge. A single-entry card does not require an identification photo. A multiple entry card requires three regular passport photos. You are not permitted to work in Mexico on a tourist card, and that includes flying passengers for hire or compensation in Mexico.

As this book is written, smallpox vaccination certificates are not

Fig. 2-8. World War II radial engine and propeller are mounted in front of the terminal building at the Ensenada, B.C. airport. Mexican flag flies proudly in the background.

required for travelers from the U.S., provided that they have not visited any other countries within 14 days prior to crossing the border. Cholera and yellow fever certificates may be required for arrivals from infected areas. A typhus certificate has been recommended in the past by the U.S. Public Health Service.

Once you have your outbound flight plan, your General Declaration form and Tourist Cards for all on board properly recorded and signed by the Airport Comandante or his representative, you will be directed back to the Operations Office where the dispatcher will now sign your original copy of the flight plan and initial the copy of your General Declaration. You're on your way into Mexico.

The General Declarations form and Tourist Cards must be returned when you cross the border on your outbound flight into the U.S. The last few times the authors crossed the border at Mexicali, the Customs office was out of the General Declarations forms, and the initial Flight Plan was filled out to serve as both documents. In this instance, we kept the initial Flight Plan/Declaration throughout our stay in Mexico and turned it in as we departed Mexico as the General Declaration form.

ONE VISITOR'S REPORT

Everyone who takes his airplane across the border for the first time is bound to have some butterflies. Pilots approach this experience in different ways. Some join in a group flight (see Chapter 4), some go with one or two aircraft and a pilot who has been there before, and others go it alone and come out fine. Take the case of Bernie Helgesen of Elgin, Illinois, who took his venerable Cessna 170B and his Honda 90 motorcycle and toured the Yucatán Peninsula. We'll have details of Bernie's junket in Chapter 7, but his thoughts and reflections on crossing the Mexican Border for the first time fit in this chapter. Helgesen prepared this material especially for this book:

"Border crossing is easy. Before departing the U.S., I checked with Customs at Brownsville. A tip for anyone carrying any foreign things like cameras, binoculars, Honda 90s, small portable radios, etc., is to list them on a U.S. Declaration Form. These forms are available at all U.S. Customs Offices. Put down the name of the foreign equipment and all the registration numbers, and have it signed off and stamped by a U.S. Customs Official before departing the U.S. Then it's already declared and there's no question about it.

"Even down in Mexico, when officials asked what I had, I just showed them the U.S. form all filled out and there weren't any questions, especially coming back into the U.S. The Customs man in Brownsville told me that if everyone would do this, they wouldn't have any problems taking their stuff in and out of the country. It sure would help them and the travelers as well.

"From Brownsville, I filed a DVFR flight plan for crossing over the border by phone with FSS and asked what frequencies to use to talk. As soon as I was airborne, I opened my flight plan and closed it almost in the same breath, and they notified Matamoros—it's only about a ten-minute flight. As soon as I closed it, I got hold of Matamoros Tower for landing; they spoke English pretty well, as did all of the towers that I contacted. They have the accent, but they were understood. If I didn't understand, I asked again. Actually, there were no problems at all.

"Customs was very helpful. They're very formal down in Mexico. They're all officials and they want to be recognized as such. If you do so, there's no problem. They had a young fellow come out who took the forms from one Comandante to the Customs to the Tower—there are three or four people that you have to see. I got to talking with some of the people in the Tower, and when it got to be quitting time, they offered me a ride to town because I told them I was going to stay over. I had to see about getting the insurance that is required in Mexico for flying. They even made a couple of phone calls to find out where it was located. It's right up where you cross the border with your auto. You get the coverage you need. It runs about a dollar a day for the time you're in Mexico.

"I met a guy at the insurance office who ended up taking me back to the hotel. So everyone was very friendly and very helpful. I can't say a thing bad about them.

"One little tip—when you go into the offices like the Commandante's at the airport, and he's very VIP, take off your hat as you walk in. Take off your hat, be formal and friendly, and you'll get along just fine with them. They kinda like that.

"You must file a flight plan anywhere you fly in Mexico. I really don't think they go by it too much. Be sure you check in at the Airport Comandante's office when you get back and close your flight plan. Do that as one of the first things—and don't forget to take off your hat when you walk in. They'll give you a big smile for it right now" (Fig. 2-9).

A CESSNA TOWED BACK

There is no reason to have a problem with your airplane in Mexico. However, should you bend your bird south of the border, you can have it hauled back to the U.S. for repairs. The following account will give some indication of the problems encountered by David A. Bassett, Wilderville, Oregon, when he purchased a derelict Cessna 170B in Mexico and hauled it back home to rebuild. Mr. Bassett first reported this escapade in the *Cessna 170 News* and kindly rewrote it for this book:

"I first saw XB-XUZ, a 1955 Cessna 170B, serial number 26631, in early 1976 on an abandoned old airport in tropical Manzanillo, Colima, Mexico. Despite the obvious vandalism and forlorn state of repair, I could see real potential in the Doyn converted 180-hp Lycoming late model 170B.

"The plane was owned in Manzanillo by Robert Hallsey, an American and long-time Manzanillo resident who operates a Mexican *super mercado.* Mr. Hallsey, an energetic 74 years old, is now engaged in real estate activities with the assistance of his wife Carmen, a native of the area. Hallsey had used XB-XUZ as a tremendous improvement over Mexican

Fig. 2-9. Visitors who don't speak Spanish usually "get by" very well in Mexico. As shown in this Mexican hotel swimming pool sign, the English language is not always translated with 100% accuracy, unless the sign painter *really* meant, "the proper swimming clothes 'might' be worn."

highways and had kept it at the old airport, which is some miles inland, to protect the bird from the salt air—and, he thought, vandals. Unfortunately, someone stole the instrument panel, magnetos, tailwheel assembly, and seats and proceeded to rip out the headliner, slash the tires and break all the plexiglass.

"After assessing the situation and doing some mental gymnastics with an estimated value of the plane back in the states (and considering the improbability of successfully getting it there) I offered Mr. Hallsey what I thought to be a reasonable amount. He countered with seven times that much. End of negotiations!

"Nearly a year later I decided to embark on a one-month vacation, driving my Chevy Blazer south to Manzanillo to see my parents who were vacationing there. All went smoothly until I arrived at Manzanillo and, out of curiosity and love for 170s, I visited the old airport and found to my surprise that XB-XUZ was still there. It was now occupied by several rat nests and some tropical black bees, was thoroughly overgrown by weeds, stickers and thornbushes, and was generally in worse condition than it had been the year before.

"I found that Mr. Hallsey was now interested in selling because the old airport would soon be closed.

"When it appeared that Hallsey and I could come to terms, I spent the next three days attempting to locate a suitable trailer upon which to bring the bird home. Being completely unsuccessful, I decided to spend part of a day in disassembly and thought that if all else failed I would at least benefit someone else. Six hours later, under a blazing tropical sun which nearly dehydrated me beyond the limits of *cervesa* (beer) recovery, I had the thornbushes cut back, the airplane pulled out into the open, the wings ready to remove, and the tailfeathers off and safely stowed in the cabin. After exhausting all possible leads on trailers over the next three days I returned to my first idea, which was to locate some tires, put the plane back on its own feet, carefully support the wings alongside the fuselage, attach the tailwheel spring to the trailer hitch on the Blazer, and tow it home backwards. I was skeptical about the ability of the water-soaked wheel bearings to travel some 1300 miles to the border.

"After a trip to the airport at Colima, capital of the state, I came up with one 6.00-6 tire and tube and one 8.00-6 tire with a 7.00 tube. Both were very used and beyond the limits for safe service in the air but were the only tires available. New *llantas* (tires) would have to come from Guadalajara or Mexico City. I decided to try the used Goodyears on the Cleveland wheels (after much cleaning, scraping and repairing) and headed north hoping to pick up some tire and tube reinforcements at Puerto Vallarta, Mazatlán or wherever.

"Dad and I spent the next morning remounting the tires and removing

Fig. 2-10. Cessna 170B with the tailwheel tied securely to a Chevrolet Blazer prepares to head for the U.S. Border (courtesy David A. Bassett).

Fig. 2-11. Cessna 170B with wings lashed along the fuselage prepares to depart for the U.S. This is not a conventional way to haul an airplane more than 2500 miles (courtesy David A. Bassett).

the wings and carefully supporting them along the fuselage with parachute shroud, nylon rope, and other assorted materials that I had put in the Blazer before leaving home. In this condition. I lashed the tailwheel spring to the hitch and tried the whole package for towing stability on the old airstrip (Fig. 2-10). Dodging thornbushes and chuckholes provided a good test to show that not only did the plane tow extremely well backwards and appeared quite stable, but also it did not seem to unduly wear the already badly-worn tires.

"Rather pleased at this point, I set about getting the final paperwork underway with the Federal Tax Office (which controls all buying and selling) and to pay Hallsey the agreed price (Fig. 2-11).

"Leaving the next morning, I waved goodbye and began traversing the torturous winding mountain roads. They were slow going, but did have a good asphalt surface. One horrible exception were the streets of Puerto Vallarta which are cobblestoned, narrow, and full of buses and pedestrians with dogs and kids everywhere. After getting my blood pressure somewhat back to normal, I stopped at the airport just north of town in search of better tires and tubes. Try as I might, the officials seemed bent on impounding the plane and forcing me to take it out to the ramp and tie it down where planes obviously belong. Finally, after much halting explanation, I convinced the personnel there that all I really wanted were tires, which quickly brought the answer *'No hay llantas.'* Glad to be free of the hassle, I proceeded northwest to Tepic and negotiated horrendous traffic before stopping for the night.

"I departed at dawn the next morning and proceeded over mountainous roads with surfaces worse than the cobblestones of Puerto Vallarta. At Mazatlán airport (Fig. 2-12), I parked a sufficient distance outside the gates and walked in search of tires with the same negative results. Carefully examining the Goodyears, I decided they looked no worse than when I put them on and proceeded north toward Los Mochis where I planned to spend the night.

"After swallowing a helping of bad gas and water which ol' Blue had picked up north of Mazatlán, the Blazer settled down and seemed to be recovered when all of a sudden a Buick Skylark sped around me with olive drab uniforms waving from every window. I was being commandeered by the Mexican army!

"They flagged me down, insisting that I stop in the middle of the road (even though a wide spot was less than 100 feet ahead) and demanded, with rifles at the ready, all my paperwork on the airplane. Attempting to greet them with a cheerful *'Buenas tardes,'* I proceeded to convey in terrified and limited Spanish what I was trying to accomplish. They wanted to see every placard, every serial number, and every registration mark on the plane. Meanwhile, another flatbed truck of soldiers arrived, making a total of 25 men surrounding me—something like ants on a wounded beetle. They were

Fig. 2-12. Mazatlán Airport where Dave Bassett attempted to purchase used tires for his old Cessna.

Fig. 2-13. Chevrolet Blazer with dismantled Cessna 170B in tow was photographed on a Mexican highway during an epic trip back to the United States where this aircraft was restored (courtesy David A. Bassett).

not satisfied with my papers and insisted on placing Ramon, a Mexican soldier in full combat uniform complete with FN FAL rifle, in the Blazer to accompany me to their military base just north of the little town of Guamuchil. On the way to the camp, I pondered what I might have done wrong, and came up with the possibility of interstate transport without proper licenses. Other than that, I was at a loss to know what really was happening.

"As the gates of the 68th Infantry Batallion of the National Army of the Republic of Mexico swung open, I knew I had seen my last daylight as stories of guilty until proven innocent Americans in Mexican prisons filled my mind.

"After a captain escorted me to headquarters, I was somewhat relieved to find a two-star general with a pleasant attitude who appeared to be versed in the Mexican policy of encouraging tourism and treating *Norte Americanos* in a professional and polite manner. I handed the general my aircraft bill of sale, passport, tourist visa and the like. He excused himself for 15 minutes and returned with a very welcome 'okay.' I can only presume that he compared the registration and serial number with a list of stolen aircraft and not finding mine among the missing, he allowed me to proceed on my journey. The relaxed stance, attitude, and friendly smiles of the soldiers were a welcome change as I departed the base after receiving the general's blessing (Fig. 2-13).

"Badly shaken but undaunted, I proceeded north and spent the night in Los Mochis, again leaving at dawn with the border possibly within my grasp. By that evening, I had negotiated the streets of Hermosillo, Guaymas and other Mexican towns enroute to the border and the dreaded Customs where surely the defecatory material would hit the oscillating blades, as an American in Manzanillo had so vigorously assured me it would. (While in Manzanillo, every acquaintance and person to whom I explained my plans of transporting the 170 home said 'It can't be done; don't try it; it isn't worth it.' This only served to make me more determined in my desire to see a good airplane back in the air again.)

"Tension began to mount as I entered Nogales just at sundown with the traffic bumper-to-bumper, block after block. I missed the turn to the border and proceeded to negotiate downtown in a traffic jam that would put the L. A. freeways to shame.

"Finally arriving at the pearly gates, I could see the Mexican Customs officials looking over the row of cars ahead and gesturing at me. I arrived at the point of reckoning where they demanded 'papers!' The Customs Officials were obviously displeased, shaking their heads and saying 'No, no, no!' as the line of cars behind me stretched out of sight. I attempted to explain that the Comandante in Manzanillo and El General had both approved my venture. Finally, the befuddled official looked at me, looked at the airplane and the long line of cars stacked up behind me, held up the papers, handed them to me and said, 'Okay.'

"Not waiting for another answer, I jumped into the Blazer and proceeded the next 100 feet or so (which seemed like an eternity) to the United States side where a big smiling Customs official spoke *English!* He briefly checked my passport and said 'Pull up under the canopy for a declaration of merchandise.' As luck would have it, the next Customs official was interested in planes and seemed to be about my age. He greeted me with, 'Okay, what's the story?' After unfolding it, he expeditiously filled out a form indicating the return of a U.S.-made product, and I was off on a beautiful freeway to Tucson. I was very appreciative of what I had in my mother country.

"After one more roadblock on the freeway by U.S. officers asking what country I was from, I smiled and replied 'Oregon.' I really felt home free.

"The entire event was a real tribute to the integrity of the Cessna airframe as the gear box, panels, and all other sections are totally still sound, and to the Cleveland wheels and Goodyear tires which went some 4000 landings beyond their service limit."

Chapter 3

There Are Differences

The same basic "dos and don'ts" apply in Mexico flying as in the U.S., yet some pilots seem to "let their hair down" when they cross the border and are out of reach of U.S. authorities. Our recommendation is to fly the same way in Mexico that you do in the U.S.

Don't buzz, though flight at 500 feet down the broad isolated beaches does no harm and makes great sightseeing.

Don't push the weather and *don't* push darkness. *Don't* push the fuel range of your aircraft. *Don't* push yourself. If you've been partying half the night, give your system eight to twelve hours to catch up.

Don't joke about narcotics. It isn't a joking matter on either side of the border. The last thing you want on your Mexican visit is to be detained on suspicion of drugs. 'Nuff said' (Fig. 3-1).

Unlike flying in the U.S., *don't* pick up passengers along the way. Your General Declaration lists the people you should have aboard. If the number of people or the names differ, you'll have some difficult explaining to do.

Don't be impatient at any of the many ground delays. Remember, no matter how difficult it may be at that particular minute, you're a guest in a foreign country.

Do take your time and enjoy a whole new set of sights and sounds.

Mexico does not have the 122.0 "Flight Watch" system that has become so popular in the U.S. in recent years. Unicom and air-to-air frequencies—usually 122.8, 122.9 and 122.95—are used by aircraft in flight to keep up-to-date on weather along their route on busy weekends when there are a number of English-speaking pilots in the air.

Another updating system is to use a high-flying, English-speaking jet (Fig. 3-2). As an example, this system was useful to us after spending a night at the isolated strip at Santa Inez, Baja, California, where there is no way to get a weather briefing. There were remnants of a coastal fog bank when we took off at mid-morning. The decision had to be made within the first half-hour of flying whether to buck the fog and go to Tijuana or hop over the mountains and return to Mexicali. We tuned Hermosillo Center on 128.3 and waited for a Stateside voice to come on the air. It happened to be Western Airlines flight 741 from Mexico City to Los Angeles. Sure, he'd be happy to check the San Diego current weather and relay. It was good enough to swing to the west and drive on up the coastline.

We'd strongly suggest that you get the current frequency for each of the Mexican high-altitude centers before crossing into Mexico. Under deregulation, you will find a good many English-speaking crews in the air carrier cockpits who will be happy to relay weather information or other data. At the time of our visit, Hermosillo Center was 128.3, Mazatlán was 124.2 or 128.0, and Mexico City was 120.1. However, these frequencies are subject to change, so get an update.

If you should have a radio failure in flight, procedures are the same in Mexico as in the U.S. As you approach your Mexican airport, fly a normal pattern with a 45° entry leg and your landing light on. If nothing happens, fly a normal pattern with your light still on. You may wish to cheat just a little so that the light points toward the tower. Sooner or later, you'll get a green

Fig. 3-1. Piper Pawnee duster takes off from an isolated flight strip near the retirement community of Santa Rita Mountains, Tucson, Arizona. The aircraft had previously been grounded and 290 pounds of marijuana found nearby. The pilot and plane escaped after this takeoff (courtesy Dan Tortorell, staff photographer, *Tucson Daily Citizen*).

Fig. 3-2. High-flying jet transports like this AeroMexico DC-10 will frequently relay current weather information to lower flying general aviation aircraft. Flight crews of international aircraft all speak English to a greater or lesser degree (courtesy Mexican National Tourist Council).

light for landing. We had this happen on one flight inbound from La Paz to Mexicali. After clearing Mexican customs at Mexicali, we telephoned the tower on the ground and arranged for light-gun instructions for takeoff and we telephoned for Customs notification at Calexico. Calexico's airport had Unicom only at that time, so a no-radio landing was routine. We then closed our inbound border crossing flight plan on the hot line to Imperial FSS.

AIRPORT SERVICES

The services you receive at Mexican airports, at least the 48 largest, come from *Aeropuertos y Servicios Auxiliares* (ASA). They have a staff of 4000 people handling airport maintenance, tiedown areas, fueling, and maintaining emergency equipment. ASA also operates the fine new general aviation terminals in Mexico City, Acapulco, Guadalajara, Puerto Vallarta, and Mexicali (Fig. 3-3).

Non-ASA airports are privately owned and operated. You can expect anything from nothing more than a dirt strip to relatively complete airports operated by the larger resort hotels. The smaller fields may not even have tiedowns, so bring your own (Fig. 3-4). Most private fields do not have fuel; if it is available, there's a great problem getting it to your airplane, and credit cards are usually not accepted.

In 1979, NACOA, formerly in charge of fueling, merged with ASA so that a single government entity takes care of fuel and oil.

Servicios a la Navegacion en El Espacio Aereo Mexicano (SENEAM) is the government agency that provides meteorology, communications, air traffic control, and navaids. Prior to 1978, these services were a part of *Radio Aeronautica Mexicana, Inc.* As of May, 1979, in operation under SENEAM were:

- ☐ 39 Control Towers (TWR)
- ☐ 10 Approach Control Services (APP)
- ☐ 4 Area Control Centers (ACC)
- ☐ 24 NDB
- ☐ 51 VOR
- ☐ 36 DME
- ☐ 9 ILS
- ☐ 1 Enroute Radar System
- ☐ 5 Airport Surveillance Radars
- ☐ 6 Weather Information Bureaus
- ☐ 1 Weather Forecast & Analysis Center (Mexico City)
- ☐ Aeronautical Telecommunications Network

Flight Dispatch and Flight Control Service is being implemented. Assistance in flight plan preparation, weather information, current NOTAM summary, and emergency procedures will be available in the near future.

Airex, one of the FBOs at the Mexico City general aviation terminal, offers a complete list of executive services. Airex is open 24 hours a day, seven days a week including holidays, to handle transient aircraft, particularly computerized flight plans for larger corporate planes. (The authors were told that Airex offers services at other airports throughout Mexico, but we were unable to verify.) You can call Airex at 905-763-4400 for information or write at P.O. Box 6740, Mexico 6, D.F.

Beech, Cessna, and Piper also have FBO facilities on the general aviation side at Mexico City. If you're flying a Cessna, you'll probably feel more at home with the other Cessnas, and the same thing applies with Beech and Piper.

FUEL PROBLEMS

There are some of the same fueling problems in Mexico that are found in the U.S. Some airports don't have 100 octane; some don't have 80; some don't have any at all.

When George and Madge Craig of Milpitas, California made an unscheduled stop at Ixtapec because of weather, they found no fuel available at the civilian airport. George contacted the Airport Manager and was advised that the only av/gas in the area was at the military base 15 miles distant. After finding the Comandante and establishing the price, Craig went along with a Captain and two enlisted men in a jeep to take the gas back to the civilian airport. He attempted to obtain permission to use the military field

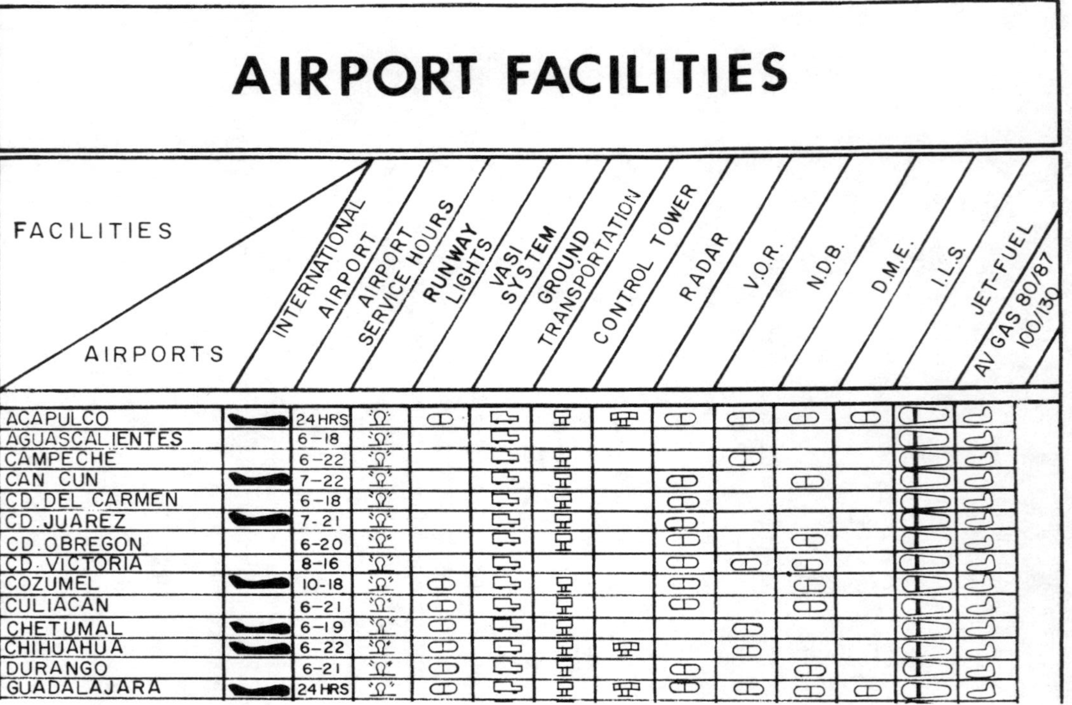

AIRPORT FACILITIES

FACILITIES / AIRPORTS	INTERNATIONAL AIRPORT	AIRPORT SERVICE HOURS	RUNWAY LIGHTS	VASI SYSTEM	GROUND TRANSPORTATION	CONTROL TOWER	RADAR	V.O.R.	N.D.B.	D.M.E.	I.L.S.	JET-FUEL	AV GAS 80/87 100/130
ACAPULCO	✓	24 HRS	✓	✓	✓	✓	✓	✓	✓	✓	✓	✓	✓
AGUASCALIENTES		6–18	✓		✓							✓	✓
CAMPECHE		6–22	✓		✓	✓			✓			✓	✓
CAN CUN	✓	7–22	✓		✓	✓		✓		✓		✓	✓
CD. DEL CARMEN		6–18	✓		✓	✓		✓				✓	✓
CD. JUAREZ	✓	7-21	✓		✓	✓		✓				✓	✓
CD. OBREGON		6–20	✓		✓	✓		✓		✓		✓	✓
CD. VICTORIA		8–16	✓		✓			✓	✓	✓		✓	✓
COZUMEL	✓	10-18	✓	✓	✓	✓		✓		✓		✓	✓
CULIACAN		6–21	✓	✓	✓	✓		✓		✓		✓	✓
CHETUMAL	✓	6–19	✓	✓	✓	✓			✓			✓	✓
CHIHUAHUA	✓	6–22	✓	✓	✓	✓	✓		✓			✓	✓
DURANGO		6–21	✓	✓	✓	✓		✓		✓		✓	✓
GUADALAJARA	✓	24 HRS	✓	✓	✓	✓	✓	✓	✓	✓	✓	✓	✓

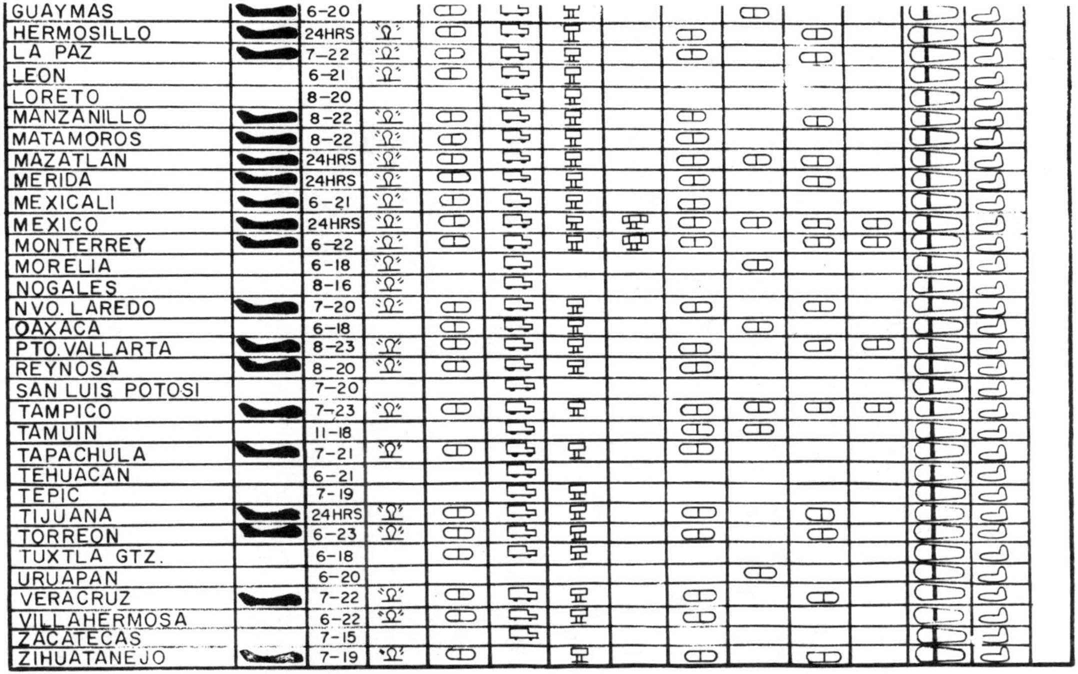

Airport	Hours
GUAYMAS	6-20
HERMOSILLO	24HRS
LA PAZ	7-22
LEON	6-21
LORETO	8-20
MANZANILLO	8-22
MATAMOROS	8-22
MAZATLAN	24HRS
MERIDA	24HRS
MEXICALI	6-21
MEXICO	24HRS
MONTERREY	6-22
MORELIA	6-18
NOGALES	8-16
NVO. LAREDO	7-20
OAXACA	6-18
PTO. VALLARTA	8-23
REYNOSA	8-20
SAN LUIS POTOSI	7-20
TAMPICO	7-23
TAMUIN	11-18
TAPACHULA	7-21
TEHUACAN	6-21
TEPIC	7-19
TIJUANA	24HRS
TORREON	6-23
TUXTLA GTZ.	6-18
URUAPAN	6-20
VERACRUZ	7-22
VILLAHERMOSA	6-22
ZACATECAS	7-15
ZIHUATANEJO	7-19

Fig. 3-3. List of ASA operated airports in Mexico. This information would be current only at the time this chart was prepared (courtesy ASA).

Fig. 3-4. Aircraft tied down beside the flight strip at Bahia de Palmas, B.C. Many small resort flight strips do not have tiedowns, so play it safe and bring your own.

that was being used at the time by civilian *fumicadores* (duster planes) but was not able to fly in. Eventually he had a half barrel of fuel in his tank and was ready to go when the weather cleared.

In back country areas where av/gas is not available, there is always the temptation to take any kind of gas you can get just to fly back to civilization. *Don't do it*! Expensive aircraft engines were designed to run on a certain grade of av/gas. Use any fuel with a lower octane and you not only blow any warranty you may have on your engine, but you also run a high risk of a forced landing in an area that could well be much worse than the locale where you originally ran low on fuel.

Some of the fuel sold in Mexico is not color coded. If the pump or the tank or the gas drum says 100 octane, then there's a good chance that's what it is.

Fuel from drums can become contaminated if the drum has been opened too long. This has happened to us only once in all of our years of flying in Mexico (Fig. 3-5). It was at Bahia de Los Angeles in Baja California. We were headed for La Paz in our C-170B before we had installed long range tanks. (Actually, it was this fueling that encouraged us to put in larger tanks.) Wind indicated that we'd be fairly high and dry arriving at Mulegé, 425 statute miles south of Mexicali, so we elected to play it safe and stop at L. A. Bay for just ten gallons of fuel. We took the fuel in the right tank, and the fuel came out of the low end of a 55-gallon drum. It was poured in out of a five-gallon "jerry can" through a chamois. We checked the wing sump after filling and the liquid that came out of the drain looked clear. We found out later that it *was* clear—it was water! (Fig. 3-6)

Many years ago, some of us were taught to fly on the theory that any engine can quit at any time. To keep this disquieting situation from happening at the worst possible time, we were taught to land and take off on our original, proven-to-be-reliable tank, and to change tanks only when a good landing spot was available. That's how it happened that we changed tanks over the isolated but usable flight strip at El Barrill, some 60 miles south of L. A. Bay. Our new 180-hp Lycoming engine sputtered in disgust and quit. We changed tanks and it picked up again. We kept switching tanks, getting as much use as we could from the contaminated tank and climbing over El Barrill. At that time, we thought we had a plugged fuel line. When we reached 8000 feet, we turned south and closely eyeballed each and every beach, dirt road, plowed field, or whatever until we'd passed the Tres Virgenes volcanos, looked at Santa Rosalia's then-deserted airport, and slid downhill into Mulegé.

When we finished draining the water from the offending tank, we gave the rest of the fuel away to eager cabbies. The next morning on the way to the airport, our cabbie's engine stalled from fuel that came from our plane. We drained that tank completely not once but *three times* in the subsequent month and were finally able to purge the water and resulting fungus only by

Fig. 3-5. Fueling facilities at the "old" airport at Bahia de Los Angeles. Structure in the foreground was once used to pen turtles that were transported live to Ensenada by truck. Fuel was brought in on the return load.

Fig. 3-6. Fueling from an old drum of gasoline with a hand pump similar to the one shown here resulted in watered av/gas in the authors' tanks. Note clothes pin holding chamois skin to the top of the funnel.

rinsing it with methanol—a non-FAA-approved fix that worked. Of course, the carburetor also had to be pulled and cleaned.

By changing from 18½ to 26-gallon fuel tanks on each wing, we now have the range to fly from Mexicali to Mulegé, and on to Loreto if we wish (490 statute miles) without fueling, but that's really too long to sit in one little airplane at one time.

Aside from the Mexicali-Mulegé hop, the only other long leg without refueling is Mexicali-Guaymas. All the other large cities in Mexico and Baja are well within 300 miles of each other.

SERVICE A WEATHER SCHEDULES

Aviation Weather Collection, Mexican Stations

ACA Acapulco, Guerrero
CNA Cananea, Sonora
CUL Culiacan, Sinaloa
CVM Ciudad Victoria, Tamaulipas
GDL Guadalajara, Jalisco
GYM Guaymas, Sonora
HMO Hermosillo, Sonora
LAP La Paz, Baja California
MEX Mexico, D.F.

MID Merida, Yucatán
MTY Monterrey, Nuevo Leon
MZT Mazatlán, Sinora
NAU Nautla, Veracruz
NOG Nogales, Sonora
PVR Puerto Vallarta, Jalisco
TAM Tampico, Tamulipas
TUX Tuxpan, Veracruz
VER Veracruz, Veracruz
DGO Durango
CUU Chihuahua
TRC Torreon
PPE Puerto Penasco, Sonoro (1400-0300)
TMN Tamuin San Louis Potosi (1500-0200)

Mexican Aviation Terminal Forecasts

MMMX-Mexico City

(FACA) Plain language significant weather forecast for all of Mexico, including lower California covering an 18-hour time period. Winds and temperatures for 41 grid points between 15 degrees north and 30 degrees north for 5, 10, 15, 20, 25, 30 and 35 thousand foot levels are also included.

(FTMX1) 24-hour terminal forecasts for:

MEX Mexico City, D.F.
VER Veracruz, Ver.
ACA Acapulco, Gro.
GDL Guadalajara, Jal.
MTY Monterrey, N.L.
TAM Tampico, Tamps.
MID Merida, Yucatán

(FTMX2) 24-hour terminal forecasts for:

TIJ Tijuana, B.C.
HMO Hermosillo, Son.
CEN Ciudad Obregón, Son.
MZT Mazatlán, Sin.
PVR Puerto Vallarta, Jal.
ZIH Zihuatanejo, Gro.

(FTMX3) 24-hour terminal forecasts for:

NOG Nogales, Son.
MXL Mexicali, B.C.
PPE Puerto Penasco, Son.
LMM Los Mochis, Sin.
CUL Culican, Sin.
LAP La Paz, B.C.
SRL Santa Rosalia, B.C.

(FTMX4) 24-hour terminal forecasts for:

CJS Ciudad Juarez, Chih.
SLP San Luis Potosi, S.L.P.

CUU	Chihuahua, Chin.	LEO	Leon, Gto.
TRC	Torreon, Coah.	DGO	Durango, Dgo.

(FTMX5) 24-hour terminal forecasts for:

NLD	Nuevo Laredo, Tamps.	TMN	Villa Tamuin, S.L.P.
MAM	Matamoros, Tamps.	TUX	Tuxpan, Ver.
CVM	Ciudad Victoria, Tamps.	MTT	Minatitlan, Ver.
CDM	Ciudad Mante, Tamps.	TGZ	Tuxtla Gutierrez, Chis.

(FTMX6) 24-hour terminal forecasts for:

CZM	Cozumel, Q.R.	VSA	Villa Hermosa, Tab.
CTM	Chotumal, Q.R.	OAX	Oaxaca, Oax.
CME	Ciudad Carmen, Camp.	IZT	Iztepec, Oax.
CPE	Campeche, Camp.	TAP	Tapachula, Chis.

SEARCH AND RESCUE

One of the things you soon learn about flying in Mexico (and it comes as a bit of a shock the first couple of times you encounter it) is just how a flight plan works—and how it doesn't! You are required to file a flight plan on every trip, but that doesn't mean that you can close it. Flying from Tijuana or Mexicali, for example, into the back country of Baja, you can land at a dozen small airports in a dozen days and have no opportunity to close your flight plan. But that's okay; nobody was expecting you and nobody will be looking for you. Except between larger towns, there is no flight plan transmittal and no search and rescue activity as such.

Many areas close to the U.S Border have worked out search and rescue procedures with their counterparts across the border (Fig. 3-7). One such organization is the San Diego County Sheriff's Department, which began in 1967 to aid missing tourists (both pilots and drivers) who became lost in Baja California. When a plane or vehicle was overdue, friends or relatives were urged to contact:

In the United States

- ☐ Any Mountain Rescue Team in California
- ☐ San Diego County Sheriff, CA, (714) 236-2113
- ☐ Imperial County Sheriff, CA, (714) 352-3111
- ☐ U.S. Coast Guard Rescue Coordinator Center, Long Beach, CA, (213) 590-2225
- ☐ U.S. Air Force Coordination Center, Illinois, (800) 851-3051

In Mexico

- ☐ Mexicali Fire Department, (903) 762-2193
- ☐ U.S. Consultant Office, Tijuana, (903) 386-1001

A cooperative organization, Search and Rescue of the Californias (SAROC), has been established. On the Mexican side of the border, the *Cruz Roja*, (Red Cross) the *Bomberos* of the Mexican Fire Departments, officials of the Department of Tourism with their radio-equipped trucks, and the *Radioafficionados* (Mexican ham radio operators) with the cooperation of the Federal Judicial Police form the strong nucleus of rescue responsibility. Tradionally, the preservation of life and property in the Western U.S. is under the jurisdiction of the County Sheriff. In Mexico, this responsibility rests with the Fire Chief of the region.

Lt. Robert A. Morse of the San Diego County Sheriff's Aviation Unit explained to us that, "Because of bureaucratic red tape, often it is hours to days before U.S. Military (U.S. Coast Guard and Air Force) can cross into Mexico, even in a life-threatening incident. An international treaty forbids more than five (5) military aircraft in Mexico at any one time. In any event, the U.S. Air Attache in Mexico City must obtain each individual clearance. If the emergency arises on a weekend or after the close of Mexico City business hours, an intolerable delay occurs. In these cases the only alternative is to hire a commercial helicopter at $150 to $200 per hour. In the last two search and rescue operations, the families of the victims were able to fund the cost. However, experience has proven this is not always possible. A fund has been established to be used for helicopter rental only in cases in Mexico where a life can be saved and only if the victim or family cannot pay the cost. This fund is administered by the San Diego Mountain Rescue

Fig. 3-7. U.S. Border Patrol Super Cub is typical of the general aviation aircraft used for search and rescue efforts in Mexico.

Fig. 3-8. Cessna 172 flies down a particularly barren portion of the Baja California Peninsula. Search and rescue activities in terrain like this are extremely difficult.

Team, Inc., and donations for Mexico searches may be sent to their address: P.O. Box 267, La Jolla, CA 92037, with a notation indicating it's for SAROC on the check (Fig. 3-18).

"In the activity area, this group averages about 20 to 30 SAR operations a year in Baja and they run the gamut of lost hikers and fishing parties to aircraft. In the aviation area we have had no failures in that all downed aircraft have been located. There have been several very successful operations as the result of ELT installations, including one that very clearly saved the lives of four people who had walked away but were headed for certain death from exposure. Almost all of our operations are textbook examples of multi-agency cooperation, sometimes amounting to over twenty different and distinct groups, both Mexican and American, civilian and governmental."

Dr. Arthur W. Feldman, former U.S. Consul to Mexicali, was one of the original organizers of the Search and Rescue of the Californias. As a career State Department Officer from 1942 to 1972. Dr. Feldman had already had an interesting career spanning three continents when he was first assigned to the Consulate in Mexicali as principal officer in 1964.

One of his first actions in the Consulate was to volunteer as the guide for a U.S. government helicopter involved in a search for a small child in the Laguna Hansen area of Baja California. Unfortunately, the helicopter crashed in the rough air and extreme weather conditions, although there were no injuries, this event triggered a consuming interest on Dr.

Feldman's part in all Search and Rescue related activities. To a large extent, he is responsible for the present form of the organization of the Search and Rescue of the Californias.

Similar search and rescue organizations exist in other areas along the Mexican Border. Your FAA Flight Service Station in that area should be able to provide local information (Fig. 3-9).

One of your better insurance policies on a Mexican trip is a flight itinerary left with someone in the U.S. A friend, relative, or business associate who knows who you are, what you are flying, where you plan to go and when you expect to return will be able to alert the FAA should you become overdue.

When you are flying into larger airports—those where a flight plan can be closed and re-opened—it is a relatively simple matter for an FAA station to run a telephonic search. When you depart from these larger airports to visit fishing camps or villages, the problem of search and rescue becomes more complex. Thus a detailed itinerary relayed to the FAA by some interested person in the U.S. is most valuable.

ELTs (crash locator beacons) are useful only if someone is hunting for you within the time span that the batteries operate. Several years ago when ELTs were first required, we talked with one pilot of a Bonanza at Mulegé in Baja California who was blithely planning a flight to an isolated dry lake on the west coast of the Peninsula without having advised anyone where he

Fig. 3-9. Piper Cherokee flies down the coastline of Baja California. Several areas of fairly wide beach are available for emergency landings, but it can be a very long walk back to any form of civilization.

was going. "Oh, I have an ELT on board. If I get in trouble, it will activate or I can turn it on and someone will find me," was his reasoning. *Don't you believe it!* ELTs are next-to-useless in Mexico unless you become overdue and the officials know about it. When an aircraft is known to be missing in a certain area, air carriers and other aircraft transiting the area may be requested to monitor the 121.5 emergency frequency, but don't really count on it. This same situation, of course, applies in isolated areas of the U.S. where flight between outlying airports are made without the capability of opening or closing a U.S. flight plan.

A number of survival booklets have been printed. The USAF and the Canadian Air Force manuals and *Private Pilot's Survival Manual,* (TAB Book No. 2261) are excellent guides. Many pilots planning Mexican flights or junkets into any back country area will carry one of these publications. In cases of after-crash shock or exposure, it has been proven that using proven survival techniques "by the book" will increase your chances of staying alive.

FROM THE BACK SEAT

Our perenniel backseat companions on Mexican junkets have been relatives Paul and June Crawford of Santa Barbara, California. They're retired and dearly enjoy traveling. They now have the leisure time to take off and go on short notice. We asked them to write their reactions to several of our mutual Mexican flights. Here's what they had to say:

"Having traveled with Don and Julia before, we all quickly fall into a routine that is comfortable to all of us. Don is usually the pilot and does all the airport red tape, gas, plane inspection, etc.; Julia is co-pilot plus accountant and treasurer; June is the researcher of hotels and restaurants; I'm the linquist, baggage man, plane loader, and general handyman. Maybe the key to having fun on such a trip is that no one hassles anybody about the job each is doing. Don and Julia make all the decisions in the plane. Hotels and restaurants we settle easily with no argument. We found that all hotels and restaurants listed in the auto club tour guide were good. The one star places were reasonable, colorful and clean. They didn't have air conditioning, but they all had large fans that in most cases were better than the air conditioning systems that we did encounter. Sometime before landing, June would mark two or three hotels, preferably near the plaza. Close to the plaza means that you are in walking distance of the restaurants and the shops and village activities without taxies.

"We like and enjoy Guaymas. The Playa de Cortez is a large spacious resort hotel with not much beach, but a large, lovely swimming pool. We "taxied" (by car, not by plane) into town, walked around the plaza of the Three Presidents, visited the cathedral where a *very* elegant wedding was about to take place, and ate at a nice restaurant near the harbor. The airport is small, clean, colorful and efficient. We've never been delayed by red tape at this facility.

"In general, you will like the cities or towns that provide the things you are looking for. Mazatlán, Manzanillo, and Acapulco are big and busy. If you are looking for nightlife, it is there. Puerto Vallarta, Guaymas, Zihuatanejo, San Blas, and Punto Peñasco are the smaller and slower moving towns, with the last two rolling up the streets at dusk. In all of them the fishing is very good. In the first three, there is a choice of in-town hotels or resort hotels outside of the main part of town. The resort hotels are complete in themselves with beaches, swimming pools, breakfast, lunch and dinner areas, bars, shops and dancing, plus fishing and tour arrangements in beautiful, elegant, tropical surroundings.

"Knowing Spanish is not necessary, particularly at airports and in the cities. However, being able to communicate with taxi drivers, waiters, soldiers, or police around the airports is a lot of fun (Fig. 3-10). There is also an extra dividend in the friendliness and warmth it generates. While driving from the airport in Zihuatanejo—some 28 kilometers—Don told the driver of the airport van that we would love some cold beer. He stopped the van almost immediately in the middle of nowhere. There was a thicket about 50 yards off the road. In it we could barely see a roof of thatch. He took some money from Don and ran off into the brush. Quickly he returned with five cans of cold beer. It was very hot and muggy so the beer was a blessing. We carefully put the empty cans on the seat beside the driver and asked him to dispose of them. He did! He threw them out his open window onto the road. He saw that it startled us so he laughed and shrugged his shoulders. This is pretty much the common attitude towards littering.

"June's notes say that you should remember to bring hot weather clothes, cotton tops and skirts, cool nightie and cool, comfortable walking shoes. Also some insect repellant and washcloths—*faciales*. The only hotel that supplied one was the Oceano in Puerto Vallarta.

"In small plane flying, the schedule is not fixed. It is frequently altered because of delays—sometimes your fault. Maybe you had a good one the night before and you just didn't want to get up that early, or the man with the key to the gas pump was taking his siesta, or the man who okays the flight plan is off and his substitute took half an hour to do a five-minute job. So we always carry snacks in the plane. Crackers, tube cheeses to squeeze on the crackers, potted meat in an easy-opening can, candy bars, and some kind of pemmican or jerky are just some of the items. Take things you like that won't spoil if you leave them in the plane. We also carry something to drink, usually mineral water or soda pop. It's nice to have around when you take off at 11 a.m. instead of the 9:30 a.m. that you planned on. If you are still up in the air at 2 or 2:30, you get kind of hungry, which usually leads to crossness and irritating remarks, which doesn't make the atmosphere in the cabin comfortable.

"About the water—which means we're talking about getting the trots, Montezuma's Revenge, or whatever you call it—we have had no trouble. We ask at the hotel desk if the water is purified. Most said yes. Others supplied purified water bottles in each hallway. We always drank water

Fig. 3-10. Paul Crawford talks with Mexican soldiers prior to takeoff from San Blas. The entire conversation was in Spanish.

served at hotels and ate the salads. Occasionally, in unlisted restaurants, we would avoid the green stuff, but not the water. If the water worries you, you can always order mineral water or beer. To me there is a difference in the flavors of the various brands of beer that is much more distinct than in the U.S. Try them all until you find the ones you like the best—they are all good!

"The attitude of the guides and hotel people in Mexico is that we are kind of silly to worry so much about the water. They want you healthy and active so you can go out and spend your money and enjoy everything, so they are very careful about the water and the raw vegetables (Fig. 3-11). In all of our trips to Mexico, Central America and South America, ranging from three to thirty days, I was slightly ill only one evening in Los Mochis. I took two Lomotil pills. The others left me and went out to dinner. When they returned, I was better and in the morning back to normal. *Any* change of schedule and diet can cause the body to make adjustments. On a vacation, overeating and over-drinking plus too much sun and exercise can cause many of the troubles that are blamed on the water. We never buy food from vendors on the streets or in the marketplaces, no matter how good some of it looks, although I have seen many of the younger tourists buying and eating from the food stalls. No doubt they blamed any problems they had on the water.

"Don and Julia plan each stage so that flight time comes out between two and three hours. This makes it more comfortable on the bottom and the bladder, which is important if you want to enjoy the scenery. They fly with the passengers in mind. If it is hot, we go up to 5000—6000 feet; otherwise we mostly fly at 1500—2500 feet. With a nice long secluded beach we go down to 500 feet (Fig. 3-12). It's a nice change of pace that you don't get on an airliner. The highest we went in all of Mexico was almost 10,000 feet when we crossed over the Sea of Cortes to Baja. Everything shrunk in size, like looking through the wrong end of the telescope.

"By flying comfortably low there is something to watch all the time. The dunes in the Gran Desierto are impressive and interesting; the colors and shapes are fascinating. Other stretches worth seeing along the coast are the approach to San Carlos and Guaymas from the north (which is spectacularly rough with tiny white crescent beaches, white waves, and the intensely colorful greens and blues of the sea) and the approach from the south to Los Mochis over Topolobampo. This is a green delta that has a most interesting combination of swamp, rocks, pools and channels.

"Flying low over Kino, you can see the beach houses. There is an incredible variety and astonishing differences in architecture and landscaping. Some were very lovely and some just bizarre. In the more mountainous areas along the coast that are close to the cities are the luxurious mansions with tennis courts, swimming pools, guest houses, and beauti-

Fig. 3-11. Ornate hotels like the Fiesta Americana in Puerto Vallarta have dependably good food and water.

Fig. 3-12. Small resort beaches like this one just south of Mazatlán are much more colorful when seen from 500 feet than at so high an altitude that all detail is lost.

Fig. 3-13. Rugged mountains can be seen close at hand, yet the safety of flat plains is usually nearby. Shown here is a section of barren mountains of the San Pedro Matir Range in Baja California.

fully landscaped grounds. There are also many interesting houses that are perched right on the edges of cliffs—some so close it would make you dizzy to look out the seaward window.

"Much of the northwestern part of Mexico is desert similar to the southwestern deserts of the U.S. This makes the areas above mentioned contrast sharply. Baja is a contrast, also. The topography is much more violent (Fig. 3-13). Much more strata is visible and the colors in the bare places are much brighter. The changes from a semi-flat, dull grey plateau to a steep-sided, colorful arroyo, to a wide flat river bed and back again are abrupt, but interesting.

"Another way in which Don keeps the passengers comfortable is by flying higher if it gets hot. If it's bumpy, he looks for a less bumpy level; if it's cloudy, he goes down or outside, etc. On our latest trip, we all had earphones and a voice-activated Sigtronics intercom. It was interesting to hear what was going on although about half was completely incomprehensible aviation jargon. The conversations between planes were fun to hear. Conversation among the passengers was carried on without getting sore throats from shouting. We planned hotels and kept track of where we were (I use an AAA road map) with the headsets and microphones.

"I'm sure many friends in our age bracket wonder how we dare go off flying in a small plane. In a Christmas card this year, long-time friends wrote, "What adventurers you are—flying to Mexico and on to Lima, Peru, in a single-engine Cessna!" Guess they think we're crazy, but, we've loved every flying trip."

Chapter 4
In Years Past

Our first exposure to Mexico and its airports was way, way back in 1949, when Ryan was building the Navion in San Diego, California. They had a brand new model, N9416K, ready to deliver to a new dealer in Santiago, Chile. We volunteered.

After cleaning up a ten-point list of "squawks," we departed the factory in San Diego for Hermosillo. In those days, the airports were small, oiled strips relatively close to town. Lightplanes were a rarity and line service was an adventure. The Navion had good fuel range and we went nonstop in 4¼ hours from Hermosillo to Mazatlán, then on to Morelia and Mexico City. My logbook remarks include hazy weather, landing gear stuck, dirty gas, and poor gas (Fig. 4-1).

Navaids were the old four-course ranges and ADF homers. Towers were on a hit-and-miss basis. Weather reports were sketchy or not available. Paperwork was frequently fouled up and "gratitudes" were the order of the day.

From Mexico City, we went to the then-small town of Acapulco and used the picturesque, palm-lined airport at Pie de la Questa. A tropical storm blew up that night and we took a cab out to the field to doublecheck our tiedowns in a driving rain (Figs. 4-2, 4-3).

The hop down the coastline to Ixtapec was over country reported to still contain headhunters. It certainly didn't contain much else.

At Ixtapec, the trip really got interesting. After checking the oil, we noted a ⅝-inch chrome moly throughbolt hanging out the side of the 225-hp Continental engine. We asked a crew of Mexican Army mechanics who were working on a C-47 nearby to assess our problem. None of them had ever worked on a "flat six" and we were on our own.

Fig. 4-1. For the history buffs, here's the Mexico City International Airport in 1949. This early photo was taken from the author's Navion during an abortive delivery flight to Santiago, Chile. The modern airport is in the same location, but this airport, like other international fields, has grown immensely in 30 years.

Fig. 4-2. 1949 Ryan Navion beside palm trees at Pie de la Questa at Acapulco. The Pie de la Questa airport is still used for limited military flying.

Fig. 4-3. Mexicàn airlines parked at Pie de la Questa Airport in 1949 included a Boeing 247, a 12-passenger transport designed in 1932; an Avro Anson British WW II twin-engine trainer converted to passenger use, and the ubiquitous DC-3 in the background.

We removed the bolt and ran the engine—no problem, no leaks, no nothing. Rather than send to Mexico City for a lightplane mechanic and parts, I elected to file two flight plans (neither arrived) and fly up the middle of the Pan American Highway back to Mexico City. In hindsight, this was a mistake in judgment, but I got away with it. I was much younger and more adventuresome at the time.

It took the better part of a week for the Mexico City Navion dealer to get parts and repair the damage to the engine. The trouble stemmed from loose pile nuts installed at the factory that allowed the #1 cylinder to begin to work loose. The throughbolt was cotter-keyed and would not work loose, so it eventually broke over reported headhunter country (Fig. 4-4).

It seemed the better part of valor to return the Navion to the factory for a thorough check before sending it to Chile, where it would be the second Navion in that country. Only when we crossed the border at Brownsville, Texas, did we find out that this ship had been exported with whitewashed U.S. registry numbers. The paper work was ferocious, both in Brownsville and later in San Diego. We picked up the entire tab for this abortive ferry trip—$800 in 1949 dollars—and were happy to get out of this fiasco with no more serious difficulties. Later, Berni Dardel, a young Swiss pilot, completed the delivery flight with enough hairy adventures to fill a number of pages of an aviation magazine.

CARERRA PANAMERICANA

Back in the mid 1950s, the Mexican Government set out to prove in a most dramatic way that they had a paved highway from border to border. Some of us helped prove that they also had a viable aviation network.

The Government sponsored an annual road race in the years 1950 through 1955. The Pan American Road Race piqued interest in the sporting blood of many continents as international drivers competed in the best of road-racing and stock cars over the 1912 miles between Tuxtla Gutiérrez near the Guatamalan Border and Ciudad Juárez adjoining El Paso to the north. During those five years, a total of 87 spectators and several of the world's top racing drivers were killed.

In 1953 and 1954, I flew support for the Porsche team on this competition in the Fletcher Aviation "jet cooled" Navion. At that time, Fletcher had a contract to build Porsche engines in the U.S. and it was logical to help their competition racing team. I was a combination test pilot, executive pilot, and public relations representative for the company (Fig. 4-5).

On the 1953 race, I first drove a company station wagon from Los Angeles to Laredo, Texas, and then on to Mexico City. At Mexico City, Porsche driver Hans Hermann then drove the stock Plymouth wagon to the starting line at Tuxtla Gutiérrez with me aboard as an unwilling passenger. Undoubtedly Hermann was a great international race driver, but after watching those canyons slide by the station wagon's rear wheels—250 feet down with no guard rail—I went back to Mexico City on the airline and

Fig. 4-4. Engine is checked on the then-new Navion at the old Mazatlán terminal en route to southern Mexico. Later in this trip, at Ixtapec, a broken through bolt was discovered. The aircraft was repaired and returned to the U.S

Fig. 4-5. Porsche racing team with twin sports cars sponsored by Fletcher Aviation Corp. Famed drivers of the mid-1950s in this photograph included Carl Kling and Hans Hermann. The author is at the far left in this photo taken prior to the start of this gruelling five-day race.

returned to the factory to pick up our Navion and flight mechanic Bob Collins. Collins later drove the station wagon back to California.

On the first of these two trips, Collins and I spent our first short night curled up under the wings in a hangar at Blythe, California, and crossed the border the next day at Columbus, New Mexico. In those days you could arrange in advance for Mexican customs personnel to drive the five miles from the border town of Las Palomas and fill out all the border-crossing paperwork. The fee on a Sunday was only $11 U.S. We spent the following day with stops at Chihuahua City and Torreón with an overnight stop at Torreón.

In an effort to make up time and keep an appointment the next noon in Mexico City, we took off with full tanks at the crack of dawn for San Luis Potosí. All the airline offices on the field were closed down at this hour, so we headed south without a weather briefing.

It didn't work. An hour and a half out we ran into a solid layer of ground fog that stretched as far as we could see ahead. "The map showed" (famous last words) a good emergency field called LaColorada, 6900 feet long at 6240 feet altitude. On the map, this looked like a good place to land and sweat out the weather. When we finally found the field, the runways were all eroded. Horses grazed over the entire area, what had once been an administration building was in ruins, and there wasn't a sign of habitation within miles.

We made a 180° turn and headed back. Halfway back to Torreón we flew over a broad dry lake near the village of Symon along the railroad. It seemed a good place to check for gas, so we dragged the area carefully and landed (Fig. 4-6). The town turned out to see us. No one spoke English but, with a questionable background in Spanish, I was able to determine they had no gas. No airplane had ever landed there before, and nearly everyone was getting his first close look at a plane.

We took off and went the rest of the way back to Torreón. After three hours in the air, we were zero miles ahead. By this time, the airport was awake and we got San Luis Potosí and Mexico City weather, which was broken to scattered.

The map shows considerable civilization along the airway into San Luis Potosí, but it's no place for a single-engine airplane. What shows as roads on the map are infrequently traveled cow paths. "Villages" are two or three mud huts. The country is all high (nearly 7000 feet) and covered with enough rocks and cactus to make a forced landing a sure crack-up.

There was a long, quiet sigh in our cockpit when San Luis Potosí came into sight. The LF radio range was on at the time and we were able to double-check our navigation for one of the few legs of the flight. The hop on into Mexico City is over fairly populated country and presented no problem except that it takes at least 10,000 to 12,000 feet to thread your way around the towering peaks.

Fig. 4-6. Navion similar to the Fletcher Aviation aircraft is shown being refueled by the author on a dry lake in Northern Mexico. This photo, taken in 1954, shows fuel going from a five-gallon can to the wing tank of the Navion.

Fig. 4-7. Fletcher Aviation's "Jet Cooled" Navion is shown on one corner of the grassy airport at Tuxtla Gutiérrez with one of the two Porsche racing cars in the foreground. Fletcher mechanic Bob Collins is standing on the wing and Porsche team racing captain Baron Huschke von Hanstein is leaning on the car.

I picked up Porsche racing team director Baron Huschke von Hanstein and Bob Collins picked up our badly battered Plymouth sedan. We both headed for the race start in Tuxtla Gutiérrez (Fig. 4-7).

Since the Mexico City-Oaxaca-Tuxtla flight is on the road-race course, the racing director was interested in a bird's-eye view of the highway.

It takes a fast-climbing airplane to keep up with the highway south from Mexico City. To begin with, ground elevation at Mexico City is 7347 feet and a fully loaded airplane takes quite a roll on takeoff. Our "jet-cooled" Navion took the climb in stride with a wide-open throttle all the way up. The road goes through a pass over 10,000 feet above sea level within 30 miles of the Capital, and we had more than enough altitude to clear the pine-covered ridges.

Off to the south lie the eternally snow-capped peaks of Ixtaccíhuatl and Popocatépetl, the latter at 17,883 feet. The highway, after touching at Puebla, where there is a fine paved field, curves westward over extremely rough country. We chose to fly down the railroad past the town of Tehuacán with its Garci-Crespo airport (all 2800 feet of it at 4950 foot elevation) and then on down the railroad into Oaxaca (Figs. 4-8, 4-9).

Oaxaca (pronouced "Wahaka") has a quaint one-way airport. Regardless of wind, you land uphill and take off downhill. The strip is large enough for DC-3's and CMA operates two trips a day into the town. There is 2600

Fig. 4-8. Stock racing cars cross the finish line of the first day's competition at Oaxaca. Spectators crowded the sides of the highway that also served as a race course. When cars went off the road, many spectators were killed.

Fig. 4-9. Small sports car covers the approach to a bridge near Tehuantepec during the road race. Note that this viaduct has no guard rails in the mid 1950s.

feet of usable runway at 5158 feet. The strip is gravel and has large patches of puncture weed along the borders of the runway. Fuel and weather information are available until 2:00 p.m. daily, when the CMA crew goes home after their last flight.

Oaxaca is near the Gulf of Tehuantepec on the Pacific Ocean and thunderheads build up nearly every afternoon. Early morning flights are much more comfortable and the bulk of the schedule Mexican Airline flights are made in the morning. Incidentally, these flights are daylight, contact operation only. There are insufficient radio aids outside of the capital to try to cope with instrument weather.

From Oaxaca south, the airway follows the Pan American highway over diminishing mountain ranges to the town of Tehuantepec on the Pacific Ocean. In this town, five native spectators were killed one year when race driver Mickey Thompson of Alhambra came up over a rise in the road to find the highway crowded with Indians watching a race car that had just missed a turn and sailed 80 feet into the river bottom. Mickey swung his Ford off the road at the spot where the crowd looked thinnest and went into the river bottom himself. He was uninjured, thanks to safety belt and shoulder harness, but five spectators were not so fortunate (Figs. 4-10, 4-11).

The airport near Tehuantepec is 10,000 feet long and paved the whole way. This strip was built originally by the U.S. Air Force as a fueling spot on the way to Panama. Before each scheduled airline landing, a station wagon drives the length of the runway to shoo the cows off the flight strip.

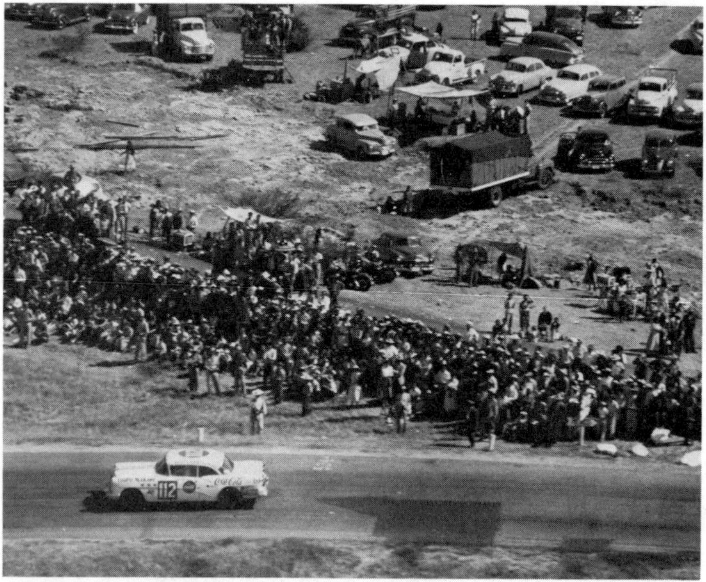

Fig. 4-10. Spectators jam the side of the paved highway during the running of the Mexican Road race in the mid 1950s. Photo taken from a low-flying Navion.

Fig. 4-11. Veteran racing driver Billy Vukovich missed a turn in Mexico and rolled this Lincoln out of contention during the road race. He was later killed at Indianapolis.

DC-3s land and taxi into the terminal at midpoint on the runway, and then take off again from midpoint.

The road south of Tehuantepec is straight for 40 miles and then winds up through the Sierra Madre Mountains into the picturesque valley surrounding Tuxtla. The airport here is an experience, particularly in larger aircraft. This field, like Oaxaca, is listed by CMA as one where the copilot is not permitted to make the landing, although the approaches are good and the surface of the runway is fairly smooth.

On race day, this airport is crowded. One year a DC-3 carried many of the Lincoln mechanics. Two press DC-3s and several extra-section airliners were present, as well as a handful of smaller planes chartered by news syndicates.

While Collins followed the day-by-day progress of the race on the ground and picking up used tires, spare parts and whatever, I took off with a variety of German mechanics. The language problem was rough since I didn't speak German and the mechanics spoke no English. We wound our way all across Mexico with sign language and an occasional "nein" and "kaput."

Bizarre rules for this race called for only a three-hour period for maintenance each day before the cars were put into impound. This was to make sure that the larger, better-financed teams would not spend all night rebuilding the racers, changing engines, transmissions, brakes, etc., each night. In previous competitions the mechanics had tuned the cars the night before, then driven all night to arrive at the finish line, leaving their cars to be started and run-up without mechanical back-up. Thus, the airplane was used to put the mechanics in two places where they were needed each day.

My logbook for that race shows a two-hour flight in the Navion from Tuxtla Gutiérrez to Oaxaca the first day, then 1:55 to Mexico City. The third day was 4:45 with a fuel stop at San Luis Potosí before arriving at the 6300-foot-high airport at Durango where the 5200-foot airport was then liberally strewn with boulders. The flight to Chihuahua City was 2:55 and downhill all the way with an additional 1:50 to the finish line at Ciudad Juárez the next day. On this final day, the paved Mexican highway was almost straight and some of the imported race cars were able to pass the Navion in flight.

The next year was different. This time I went alone, ostensibly because the station wagon was deemed unnecessary. However, that was a long trip to make alone and, while not advising company president Wendell Fletcher, I landed across town and picked up Dr. Paul MacCready, famed meteorologist for whom I'd flown a couple of years previously on two cumulus cloud research projects. Readers may remember that Dr. MacCready later (in 1977) achieved worldwide fame as designer of the man-powered Gossamer Condor, the English-Channel-crossing Gossamer Albatross in 1979, and subsequently, the sun-powered Solar Challenger. Dr. MacCready was even then a noted meteorologist, physicist, and international soaring champion. We had an interesting trip and I learned many niceties of flying weather in a lightplane.

We crossed into Mexico at Nogales, Sonora, studied weather formations in flight, and spent our first night at Mazatlán. While the two of us were walking down the paved sidewalk next to the bay, I brushed my pants leg against the cement and felt a sharp, stinging bite. I jumped and the two of us saw a small scorpion scurrying away. The leg burned vividly, so we returned to our hotel and asked the desk clerk what to do about scorpion bites.

We were referred to a medical doctor. The cab ride went down a series of dark, dusty streets and stopped at an ill-lit building. The doctor took us to his office where the symbol of medicine in Mexico—a skull with a hinged jaw—stared across the table at us. I was scared! I explained the problem in poor-but-usable taxicab Spanish and cautioned that I was allergic to horse serum, then used for tetanus injections. The doctor nodded, wrote out a prescription, and told our cabbie where to find it. We soon returned and the doctor gave the injection with one of the biggest needles it has ever been my misfortune to face. No U.S. Army physician ever had anything bigger. In the event, I survived and "sweat out" the next ten days before being sure I wouldn't have a violent reaction to the toxin.

We continued to Mexico City after loafing for a day and then on to Tuxtla-Gutiérrez, where we arranged for a ride back to the U.S. border for Dr. MacCready. He was to return in a white Jaguar with California plates and took off for Oaxaca the evening before the race started. That night the wire services carried the following report, in part: "Two Americans were killed near Tehuantepec as they drove from the start of the Pan American Race. They were in a Jaguar believed to be the service car for an entry in the small sports class of the race . . . "

It wasn't until two days later in Mexico City that we were certain that there had been two white Jaguars in the area and that Dr. MacCready was alive and well. There were some frantic, serious phone calls to his parents in Massachusetts.

The race itself was pretty much a carbon copy of the contest the year before except that we stopped for fuel at Leon and Parral out of Mexico City. At the start of the last day, team manager von Hanstein decided that neither of our two "Fletcher Aviation" team Porsches could make an overall win, so he arranged for both cars to cross the finish line near the Juarez International Airport riding abreast (Fig. 4-12). The ensuing photos made wire services all over the world, particularly since the diminutive Porsche speedsters finished third and fourth overall, competing with much larger Ferraris, Lancias and the cream of international sports cars.

In its five short years, this race had been called by Italian winning driver Umberto Maglioli as having "some of the worst roads in the world." Today, this highway and its adjoining airports provide excellent service to

Fig. 4-12. Porsche team cars cross the finish line at Ciudad Juarez in a dead heat to finish three and four overall in the international competition. This side-by-side finish had been planned in advance by team director Baron Huschke von Hanstein. Note how close the spectators are to the cars that were traveling well in excess of 120 mph.

Fig. 4-13. Writer/editor/race driver Marvin Patchen, left, and his co-driver Dave Davison, pose with their two-wheel drive Jeep "El Loco" (the crazy one) at an airport in Baja California. The Cessna Skylane in the background was flown in support of the racers by the author.

the highway traveler and the general aviation visitor. The Mexican Government well proved its point that both highways and airports were more than adequate for the hoards of visitors yet to come.

BAJA 1000

The Mexican population dearly loves its sports, and when the "Carerra Panamericana" was finally cancelled in the late 1950s, it was eventually replaced by an off-road race down the length of Baja California long before the new road was completed. Light aircraft were again used for support of the racers. Thus it was 15 years later that I again found myself flying for the racing cars in Mexico. (Note: The spelling of Santa Ynez has been changed to Santa Inés today.)

One of the most interesting of spectator sports is the grueling Baja 1000 race. In the days before the main road down the Peninsula was paved, this was perhaps the wildest of all off-road competitions. It was also really something to watch from the air.

Marvin Patchen (Fig. 4-13), who developed the very successful *Aero* magazine and later sold it, is an off-road racing buff, as is his charming wife Aletha. Marv asked me to fly a support airplane for him during the 1969 race when both he and his wife were driving—but in separate cars. It was a race worth remembering. Here's how Patchen reported the event in *Aero* magazine:

"Twice a year, some 250 assorted passenger cars, trucks, dune buggies, motorcycles, and four-wheel drive cars make the Baja Peninsula into an 832-mile junkyard in a race from Ensenada to La Paz. It is debatable as to who are the luckiest—those who finish the race or those who don't. The road is indescribably rough and dusty and a comfortable trip down the peninsula takes ten days. Pushing, it takes five. In the race you try to make it in 24 hours—or less.

"Along with the cars, more than 100 aircraft fly aerial support, carrying mechanics, wives, spare parts, the press, and spectators from checkpoint (gas stops) to checkpoint. Strips that don't have 50 operations a year are suddenly deluged with the like amount in a matter of hours. Because the race runs throughout the night, there is a mad dash for pilots to fly as far down the course as possible before they are trapped by darkness at an unlighted field.

"This year, despite the fact that some people think I broke my head rather than my leg last Spring, when I was asked to drive a two-wheel drive Jeep on the formidable Brian Chuchua Jeep Team, I accepted.

"To add to the confusion, my wife was also entered (this was her third Baja race) driving a dune buggy (Fig. 4-14) with Janet Elliott, whose husband George was driving another Jeep. We had sort of a three-way

Fig. 4-14. The first few miles south of Ensenada were on pavement. Here Aletha Patchen and Janet Elliott speed past a Mexican passenger car near the entrance to the Ensenada Airport.

worry club where frustration is not an adequate word to describe how out of touch you feel with the world or anything else once you leave the starting line. Communications are nil and the only word you get on what's happening ahead or behind is from the pilots, who spread an equal number of false and correct rumors. There is a thin band of amateur radio communication, but because Mexican operators are required, words and numbers get lost in the translation.

"You've heard of Marlboro Country—well, Baja, during the race, is Skylane Country. If you flatten Baja's mountains, you'd almost think you were in Wichita or at a 182 owners' convention. For this year's race we chose a 182 to give us our aerial support.

"Baja is the type of place where you want everything to go right and the [182's] reliable spring steel gear couldn't be more trouble-free. Its high wing is pefect for sightseeing and photography and when properly flown, its short-field ability approaches the STOL category.

"Before I got the twin-engine bug, I owned a Skylane and I must say that the several hundred hours we spent together were all pleasant.

"I asked my friend Don Downie to fly while I played Russian roulette with the chuckholes. Don is no stranger to Skylanes or Baja.

"We picked up N70915 from Santa Monica's Gunnell Aviation, pumped up the nose gear another four inches to get more prop clearance on the dirt strips, and twenty-nine hours of flying and a week later, returned home.

"While I drove to Ensenada, Don flew to Tijuana to take care of the border formalities and then to Ensenada. The morning of the race, Don went to the airport ahead of the starting time to be prepared for takeoff as soon as the team cars had passed by (they don't close off the road to normal traffic during the race). Don's passengers were Tex Carter, a master welder, Dick White from the Dana Corporation, and Carl Jackson, who rather than riding with a co-driver was racing solo from the halfway point at El Arco to La Paz after his co-rider drove solo from Ensenada. Fortunately, the team had the first starting positions (the cars started one minute apart for 251 minutes). From Ensenada, Don flew to El Rosario, 152 miles from the start. This hilltop airport, although long in length, is short on parking, and Don's early takeoff and the Skylane's speed got them to El Rosario before the bulk of the planes converged on the strip.

"Don's preoccupation with getting to El Rosario prevented him from seeing what happened below. Because of too much speed on my part, a lousy weld on someone's part on a previously repaired front axle, and a car-gobbling chuckhole on a high-speed part of the dirt road, our Jeep became a three-wheel, three-fender vehicle (Fig. 4-15).

"Don was obligated to stay with the leading team cars and could not wait to see why I hadn't come by or why my wife was far behind schedule (they had tossed the first of many fan belts on their Corvair) and took off for the next checkpoint, Santa Ynez.

"Although the number of cars began to dwindle from breakdowns, the number of planes increased. At Rancho Santa Ynez usually one skims over

Fig. 4-15. One of the smoother portions of the highway south of Ensenada. Broken down off-road racer waves that he is okay to a low-flying support aircraft.

the ranch house for an uphill landing, but at this time there was a strong gusting crosswind that favored landing downhill. Don and the Skylane made it with no sweat, but many of the other pilots worked hard at making their Santa Ynez landing . . . some didn't even try (Fig. 4-16).

"Aircraft-wise, everything was uneventful except for a helicopter pilot knocking off the top of the rudder fin on a Cessna 411 (Fig. 4-17). To the surprise of many, the next day the bent fin was cut off and the pilot took off with no problems.

"Meanwhile, back on the road, we had found a Mexican who was an artist with a portable arc welder and he remarried our axle and tie rod. We knew our front brakes were gone, but we couldn't get the remaining rear brakes to work. Dave Davidson, my co-driver, and I continued on, hoping to find somebody who would help us with our brake problem. Naturally, on this road when you have brakes it is an experience—with no brakes, it is twice the experience!

"The airplane drivers at Santa Ynez were in the midst of many calculations trying to figure out how far down course they could fly before dark. Most pilots were rooting for their cars to arrive as they hoped to make unlighted El Arco, the halfway point, before dark. As our team cars still running were amongst the leaders, Don made it into El Arco with plenty of daylight to spare.

"Arnold Senterfitt was also flying support and as darkness approached, he used Don's portable Regency transmitter and became a tower operator calling in the planes. The pilots on the ground automatically reacted to the situation without direction or prodding and parked so that their landing lights could shine on the runway. Flares were placed at each end of the runway for the stragglers who came in after dark.

"The Skylane became a motel for Don and the reclining seats and comfortable upholstery were appreciated. El Arco is an ex-mining town in the middle of Baja. Although there is only one store in town, and most of the houses have dirt floors, the people are friendly, the winter months pleasant, the air is clear and pure and they have an 1800′ fairly smooth runway, plus a 300′ (!) STOL strip. When you think of how many towns in the U.S. do not have an airport and are burdened with polluted air and raw winters, El Arco's dirt floors don't look too bad.

"The next morning Don took off for the south with Spike Cooper, who switched race car seats with Carl Jackson. Don was unaware that I had continued with the race nor had he any idea where my wife was; however, rumor had filtered down the line that I had been in an accident. Shortly after we had broken off our axle, a dune buggy went end-over-end five times, coming to rest 100 feet from our car. Those flying over assumed that the buggy and our Jeep were involved in the same incident and rumors flew!

"At Santa Ynez we made another 2½-hour pit stop, trying to fix the brakes and welding the springs to the axle. Again we gave up on the brakes and continued on. The next morning by the time we covered the 412 miles

Fig. 4-16. The tiny airport at Santa Inés was crowded with parked aircraft early in the running of the Baja 1000. Note the wind sock at the right stretched out in a stiff wind. This airport has since been paved and provides good fueling for Baja California pilots.

Fig. 4-17. A crowd gathers as a helicopter clips the tail of a Cessna 411 at Santa Inés. The Cessna pilot merely straightened out the bent skin and flew the airplane out.

to El Arco, it was almost a ghost town; Don and most of our pit crews had left for La Paz, along with the rest of the race air force. Dave and I figured that we'd better quit while we were ahead, as there were some steep canyons ahead and in a brakeless car we would be asking for more trouble.

"What happened in the next day and a half is far too confusing and lengthy to relate. The lack of communications made everything turn out as unplanned. Consequently, I spent a day and a half with an El Arco family. Much of the time was spent sleeping which the previous 39 hours had not included.

"After delivering his passengers to La Paz and the resort Bahia de Palmas, the following day Don flew back north to El Arco to pick me up, back down to San Ignacio, back up to El Arco and then to the Bay of Los Angeles, where we refueled and spent the night—especially enjoying Mama Dias's renowned turtle lobster dinners. One of our El Arco takeoffs was at full gross, 85°, with a wind condition that was of little help. The way the 182 behaved on that takeoff was exactly why so many pilots speak respectfully of Skylanes.

"At dawn the next morning we flew north up the course to see if Dave and the race car made it back to Ensenada. The plan was a stop at Ensenada to drop off our passengers who would drive a truck down to pick up the race car if it broke down. We were worried because Ensenada was fogged in, but we lucked out and found the car in El Rosario, still able to run.

"It had now been two days since I'd seen my wife and, of course, I was curious to find out if they made it to La Paz. We zoomed to San Ignacio to find that the passenger we were to pick up had found a ride to La Paz in another plane. From San Ignacio we went to Serenidad at Mulegé for gas. Most of the aircraft were returning home that day and it was panic for many of the pilots who used 100 octane fuel, as all three resorts in Mulegé were out of high octane. We spent a lot of time on the radio on our southbound trip, advising pilots of this situation so they could throttle back to economy cruise or make alternate plans. The number of planes using 100 octane had simply overwhelmed the resorts.

"I did the piloting from San Ignacio to Mulegé to de Palmas back to La Paz and sitting behind the Skylane's panel was like putting on a comfortable pair of shoes. Everything was peaceful as we were scooting along at 158 mph TAS heading for the bay named peace (La Paz).

"At La Paz I missed my wife by about 30 minutes. She had given up on us and taken a charter flight to Ensenada. I learned that their finish was a heartbreaker, but a proof of stamina for the female side of the ledger (only 37% of the cars finished the race). After numerous breakdowns, getting lost, dense fog, stuck in the mud, they broke down only three miles from the finish line. When they finally crossed the finish line, they had been driving for 50 hours continually!

"As darkness prevented further flying for the day, we spent the night in La Paz. The only remaining member of the team left in La Paz was Brian Chuchua, a Fullerton, California Jeep dealer and owner of the seven-car team. Brian was a student pilot and Don an instructor, so it was a perfect

Fig. 4-18. Cessna Skylane used on this Baja 1000 race is shown parked in the dirt near a Baja California airport that normally would see no more than two or three aircraft a week. A Cessna 180 taildragger takes off in the background.

opportunity for Brian to log some dual cross-country time from La Paz to Fullerton (Fig. 4-18).

"I'm not trying to take anything away from Brian's piloting ability, but the combination of Downie's coaching and the way the Skylane responded made subsequent landings and takeoffs smooth and safe. For those of you who have flown to the Bay of Los Angeles, you'll have to admit that letting a student pilot make the landing is unusual since so many competent pilots make bouncy landings on the strip's humpy touchdown spot. Don, of course, was alert to take over if needed, but Brian and the Skylane made the landing look simple."

As this book on Mexico goes to print, so does Marv and Aletha Patchen's new book *Baja Adventures by Land, Air and Sea.* It is a fine new publication and would make fascinating reading for travelers who want more detail on the challenging peninsula of Baja California. In his new book, Marv gives some additional background that should be of interest to those of us who like to read about adventures but really don't want to do it in the dirt. If you want to know more about this wild and wacky world of off-road racing, there is a chapter in the book on the above race entitled "1969 Baja 1000" followed by a chapter covering the 1970 Baja 500 entitled "I Try Again." 'Nuff, said. This book is available from Baja Trail Publications, Inc., P.O. Box 6088, Huntington Beach, CA 92646.

Fig. 4-19. Volmer Jensen at the controls of his famed "Chubasco" homebuilt airplane. Wings and tail are from an Aeronca Chief or Champ. Jensen has been flying this aircraft for more than 20 years.

EXPLORING WITH A HOMEBUILT AMPHIBIAN

One of the veteran explorers of the West Coast of Mexico and Baja is Volmer Jensen. A veteran sailplane and powerplane pilot, Jensen gave up flying for ten years (1948-1958) because he was tired of "airport hopping." He built four boats, learned to skin dive, and built his prototype homebuilt amphibian "Chubasco," a 100-hp two-plane, side-by-side airplane with wings and tail from a used Aeronca Chief or Champ.

In the succeeding 20 years, Jensen has sold over 700 sets of plans for the Chubasco and nearly 200 examples are flying (Figs. 4-19 through 4-23).

One of Jensen's early international flights with his Chubasco was down the West Coast of Mexico as far as Puerto Vallarta, and this was back when Puerto Vallarta was still a small town. Jensen was accompanied on this two-week trip by Rich Eshe in June, just before the rainy season.

The pair crossed into Mexico at Mexicali, with no problems from the Mexican Government for this homebuilt amphibian. Stops were at Puerto Peñasco and Kino Bay for fuel, then Guaymas, Culiacán, Mazatlán, Tuxpan, Mexcaltitan and San Blas before reaching Puerto Vallarta (Fig. 4-21).

As Jensen described it, "Following the coast all the way down and with miles of lagoons near the ocean, we flew from two feet to 500 feet off the water, which was a 3000-mile sightseeing trip.

"Landing in jungle rivers, we chased the pink flamingos with the airplane and caught a four-foot iguana by hand. We landed at times when the weather was very hot in beautiful lagoons to go swimming and cool off for a while and then take off again."

Mexcaltitan, between Tuxpan and San Blas, is a stopover spot reserved exclusively for flying boats and seaplanes. Jensen said, "We landed

Fig. 4-20. Volmer Jensen's homebuilt amphibian is shown in the water of a sheltered bay on the west coast of Mexico. This plane can land safely in three feet of water.

Fig. 4-21. "Chubasco" with its wheels down taxies from a tiedown at Jensen's home field of San Fernando in Southern California.

in three feet of water next to the town of Mexcaltitan, which was an island in a jungle swamp, and perfectly round, looking like an Aztec calendar from the air. I am sure that we were the first airplane to land there because the whole town turned out to greet us, swarming around us in their dugout canoes. We felt like Lindbergh landing at Le Bourget, and to them it was the story of the big white bird landing among them. All we could do was walk into town after anchoring my 'Chubasco' in a foot of water, and then sit down on a bench in the plaza next to the church. The people and the children all gathered around so that all we were able to do was get back to the airplane and take off again.

"We landed in small fishing villages at times to get auto gas. Flying below Culiacán, crop dusters would come up and almost touch wing tips with us. They were curious and wanted to take a close look at our 'Sportsman.' Just above Puerto Vallarta I caught a lobster while skin diving. Since we were having lunch in 20 minutes at Vallarta, we didn't want to take the time to cook it, so after we took off we found some fellows camping at a very remote area on the beach, and we dropped the lobster. It landed close to them and we saw them pick it up and wave to us. I'm sure that they had lobster for dinner that night."

Jensen's first "international" flight, back in 1960, was with photographer John Hewett. The pair spent five days at Mulegé in Baja California, and flew down to Conception Bay each day to catch fresh lobsters. Jensen reported this first trip in the Experimental Aircraft Association

(EAA) magazine *Sport Flying*. Since one good flight report deserves another, Jensen made a second trip into Baja in search of a missing airplane. Here's how the builder/pilot described this Mexican amphibious flight in *Sport Aviation:*

"Shortly after my first Baja junket, Bob Rowley, City Engineer of Los Angeles, called on the telephone and asked, 'Are you the fellow I heard about who was down in Baja with an amphibian?'

"Cautiously, I admitted that I was and wondered what would happen next.

"How'd you like to go back down there and help me find my Stinson Station Wagon?' he asked. 'It's somewhere in the Gulf north of Santa Rosalia. I can't find anybody with a landplane who will fly lower than 2000 feet unless he uses this altitude to make an approach for a landing.'

"It seemed that Mr. Rowley, accompanied by two passengers, had experienced a complete engine failure while cruising at 8000 feet some 35 miles north of the copper mining town of Santa Rosalia. The best place to land was a dry wash a half mile from the abrupt shoreline and they put the Stinson over on its back without injury to anyone aboard.

"The trio walked out to the beach and was spotted the next morning by their friends in another airplane who had been on the same trip and backtracked them from Mulegé when they didn't show up.

Fig. 4-22. Jensen's homebuilt amphibian "on the step" just prior to takeoff. Pusher propeller is high out of the water and away from any spray.

Fig. 4-23. The trim lines of the "Chubasco" stand out against a plain hillside. Paint job for the last 20 years on this homebuilt has been turquoise and ivory.

"To shorten an expensive, time-consuming story, Mr. Rowley chartered a 60-foot Mexican cruiser to pick up the damaged Stinson. With the help of a number of Mexicans, the Stinson was righted and pushed to the shoreline where the wings were removed, and using a row boat, safely put the wings aboard.

"The fuselage was towed out to the cruiser with an empty 50 gallon fuel drum attached to each landing gear and the tail in the row boat. Just as a hoist was being attached, one of the drums slipped and the fuselage turned over and floated away.

"With wind building up and night approaching, the cruiser captain returned to Santa Rosalia, leaving the airplane upside down in the water and all you could see were two 50 gallon drums floating on top. I told Bob it would be almost impossible to find the 50 gallon drums floating around. They could be 100 miles away near Guaymas or Mazatlán, but he convinced me it was worth looking for and he was taking care of all the expenses. I knew it would be quite an adventure and agreed to go.

"That's why Bob Rowley and I returned to Baja, just four months after my original trip there. His primary interest in the fuselage, now that it had been immersed in salt water for three weeks, was to determine what had caused the original engine failure since the same engine is currently in use on a number of lightplanes and helicopters.

"I put my skin-diving gear in the back of the 'Sportsman' and Bob

Rowley and I headed for Santa Rosalia. Thanks in part to Mr. Rowley's knowledge of tides in the area, it took us only 20 minutes to spot the fuselage from the air, sunk in 30 feet of water. Darkness was approaching, so we returned to Santa Rosalia and spent the night in the old Hotel Frances. Our tab was $1.50 per person.

"Next morning, we re-spotted the wreck easily. This second trip to Baja was in the middle of December and it was cold, despite the usually mild climate found in Baja. I was wearing my pajamas under my regular clothes to keep warm in the air and it would have been impossible to do any skin-diving without a frog suit, which I didn't have. We didn't land over the sunken fuselage after respotting it the next morning, because there wasn't anything we could do about it.

"Mr. Rowley returned to Baja later and spent a day in the crow's nest of a cruiser, trying unsuccessfully to locate the fuselage which had by now probably drifted into deeper water. Thus the mystery of the engine failure was never solved."

Jensen liked his travels in Mexico so well he promptly returned to Baja, this time to the Pacific Ocean side at Scammon's Lagoon. Here's how he reported the trip in 1964, long before the paved highway changed the isolation of Baja:

"This time my companion was an ex-P-38 pilot, Richard Ostronik, a fireman in Alhambra and former employee in my Production Model Shop. Our destination was the seldom-visited area of Scammon's Lagoon, site of the largest salt mines in the world and the area where grey whales spawn annually, and where Donald Douglas of Douglas Aircraft and Dr. Paul Dudley White, famous heart specialist, made an expedition to record the heartbeat of a whale because a whale's heart is similar to the human heart.

"Here's an area of the Pacific coast that is seldom visited by outsiders. Some 800 Mexicans work in the *Exportadora de Sal, S.A.,* mines that produce 99 percent pure salt. There's a nine-month drying cycle here compared with four years for similar drying near San Francisco.

"Commercial salt operations at Scammon's Lagoon are completely modern and began in 1958. There's a fine airport, probably 4000 feet long, utilized by company DC-3s, and a large hangar, complete with heat lamps to cut down on corrosion.

"The sweeping bay here is a 'catch-all' from the Japanese current. One can find Japanese glass fish-net balls, whale bones, flotsam and jetsam of all types. On a low flight—that's normally 25 feet—we saw everything from a full-sized yellow mine (probably still highly explosive), to just about anything you can imagine the sea might wash up. It's an ideal place for beachcombing.

"Scammon's Lagoon has enough history to fill a book. A British whaler, the *Tower Castle,* was shipwrecked in the entrance in 1839. Extensive commercial fishing for whales was begun by the U.S. brigantine *Boston* in 1858, and three years later, so many whalers had visited the area that the California grey whale practically disappeared.

"There are no facilities for the tourist at Scammon's Lagoon, but salt company officials were kind enough to sell us a tank full of gas and let us throw our sleeping bags in their hangar for one night. The town has two cafes, one good and one not-so-good. As in most every other place in Baja, the menu is primarily fresh-caught seafood, usually turtle or lobster.

"We made water landings at the isolated resort of San Quintin before topping our tanks again at Los Angeles Bay for the hop over the primitive area to Scammon's Lagoon. This is certainly no place for a forced landing, but that dependable C-85 never missed a beat. Near here we found a 1957 Chevrolet with the same paint job that I used on the 'Sportsman.' The people from Detroit call it 'tropical Turquoise' and 'India Ivory'—very romantic sounding names.

"On this trip, we stopped for fuel at San Felipe on the way back. The airport looks like a road and the upper portion is soft sand. However, there's plenty of room for a lightplane, and we taxied right up to the 'gas station.' It takes a tour of the cantinas in town to find the attendant—no matter how closely you buzz town—and fuel is measured with a 10 gallon milk can and strained through a chamois skin.

"With full gross weight, we were able to get off glassy water in temperature that broiled between 90 and 100 degrees. It was never necessary to taxi in a circle and create a wake to get into the air in this high temperature.

"For those planning a flight to Baja, I might mention that there has never been any problem in taking a homebuilt plane across the border.

"The drinking water is uniformly good in Baja since there is no stagnant water in the entire peninsula, and neither any of my passengers nor I have ever experienced physical problems south of the border.

"On trips of this type, I have made salt water landings for days at a time and never washed off the salt. I keep all the cadmium-plated metal parts covered with a light coat of wheel bearing grease, but I have yet to repaint the hull or replate any of the metal fittings.

"The entire trip to Scammon's Lagoon from the San Fernando Valley Airpark near Los Angeles logged 18½ hours—and 104 gallons of gas."

Jensen's homebuilt amphibian is constructed of 1/16 to ½-inch mahogany plywood and is covered with fiberglass. The plane is 24 feet long, 8 feet high and has a wing span of 36′6″. Empty weight is 925 pounds and gross weight is 1450. Wings and horizontal tail can be removed by unscrewing four bolts each, and the aircraft can be towed home on a simple trailer. Assembly or disassembly can be accomplished in less than 30 minutes by two people. For further information, send $2.00 for a brochure to Volmer Aircraft, Box 5222, Glendale, California 91201.

Chapter 5
Group Flights

Many first-time visitors to a new location, be it Mexico or anywhere else, prefer to make their first flight with a group. The group can be a simple "buddy" system with one other plane or a pre-planned junket set up carefully by a flying club, a large FBO (fixed base operator), or one of the larger flying organizations.

Group flying has certain advantages. If someone has a problem with weather, maintenance, paperwork, or whatever, there's another airplane to call or another pilot and his crew to help as your Stateside representatives. Pre-planned group flights take all the responsibility for fueling, tiedown, cabs and motels, on-the-ground activities, banquets and whatever. It is an easy way to go. For the rugged individualist it can be a pain, but he won't stay long with any group.

SKYLARKING TO MEXICO

The Sky-Larks of Southern California is one of the oldest and best-run flying clubs in the West. They have organized many fly-ins to Mexico with great success. One year, members of the club spent one Thanksgiving in Loreto, Baja, California (Fig. 5-1). The following Thanksgiving, 20 planes of Sky-Larkers congregated at Rancho Buena Vista, near the tip of Baja (Fig. 5-2). The year after that the group headed down the West Coast for Mazatlán's carnival in mid-February. The Sky-Larks could have fun just about anywhere—and they do—but their junkets into Mexico have produced a new high level in fly-in excitement and enjoyment. A group trip into Mexico is greatly recommended, particularly for pilots who have never been there before and express some doubt regarding regulations and safety. Members of flying clubs, particularly those designated as trip

Fig. 5-1. Fly-in visitors enjoy the white beaches of Baja California on a group trip. There is safety in numbers while flying over some of the wastelands of Mexico.

Fig. 5-2. Airstrip at Rancho Buena Vista as seen on short final approach. Almost all landings are made from over the water. The flight strip is 80 feet higher at the far end which helps shorten the landing roll. Pilots with tri-geared aircraft should have their nose gear struts pumped up as high as possible to cut down on prop tip damage in the gravel. Note parked aircraft at the left of the runway.

chairmen, would do well to explore the possibilities of a group club fly-in somewhere South of the Border. Contact your nearest Mexican Tourist Bureau office for assistance and your destination town and hotel for group rates, special parties, and colorful Mexican festivities.

FLY FOR FUNSTERS

The largest and possibly best-known general aviation group flights to Mexico are planned and produced by Gil and Bob Gunnell, who operate a flying service at the Santa Monica Airport. When we say these group trips are "produced," that's the proper word. Prior to each of the 61 group flights that have logged over 8,000,000 passenger miles in 20 years, the effort is truly a production. The Gunnells travel to the resort area three months ahead of time to set up hotel and banquet reservations, make sure that Mariachi bands and fishing boats will be on schedule, plan for the personal favors to be given to each guest, and generally make the show work. When the date of the actual trip arrives, it's a pair of working Gunnells who bring their own key staff and put in some very long hours overseeing while their guests do the partying.

Gil and Bob (Fig. 5-3) are great hosts, but there's also a financial incentive. Guests pay a minimum fee to cover expenses, but the primary profit for Gunnell Aviation is the sale of high performance aircraft. Bob estimates an average of two new or slightly used aircraft are sold as the direct result of each trip because guests just don't want to make that long trip in their older, slower airplane. Besides, they'd been planning for a bigger, faster one anyhow and the trip is usually the incentive to get that sales order signed.

The first of these Mexican "Fly For Funster" trips went to Guaymas in the State of Sonora early in 1962. Gil's original announcement of trip #1 was sedate when compared with what came later. She sent out an announcement and 93 people showed up. Back in 1962, the total cost for the fun-filled weekend was $31 per person!

We saw a Gunnell production on trip #56 in 1978. On a sandy beach almost an hour's boat ride out of La Paz in Baja California, we watched three chartered fishing boats, complete with air conditioning, bar, Mariachis and Funsters, arrive (Fig. 5-4). Lo and behold, there was a full-course luncheon catered by the Hotel Presidente, time to hangar-fly, swim or snorkel, get part of a sunburn, play volleyball, or whatever. Then the group returned in the boats before dressing for a unique invitation-only dinner at Caymancito, home of the Territorial Governor. All this took place during a single day of trip #56. James Bond could have done no better.

Bob Gunnell explained the sales aspect of the Funster trips:

"A great many 'first timers' on our trips have been attracted to light airplanes because of these long weekend trips that we set up every other month. (Later the frequency was reduced as trips became even longer.) The trips bring new people into aviation; they sell new planes and re-sell larger, faster aircraft to the new pilots who want a faster cruising speed."

In the first year and a half of these tours, the Gunnells sold 16 new aircraft and picked up 30 new students—including many wives. At least a dozen pilots turned in used aircraft on larger models.

"We have one new pilot (this was in 1962), an attorney, who began flying last December after hearing about our trips. He purchased a Cessna 172, turned it in on a 210, and is now flying a 310. "Even his wife is now taking flying lessons," explained Bob Gunnell.

Detailed planning for the Funsters is the work of Gil .She's a vivacious former registered nurse whose maiden name was Gilliand. It seems that all student nurses go by their last names, so she came up as "Gil" and has been

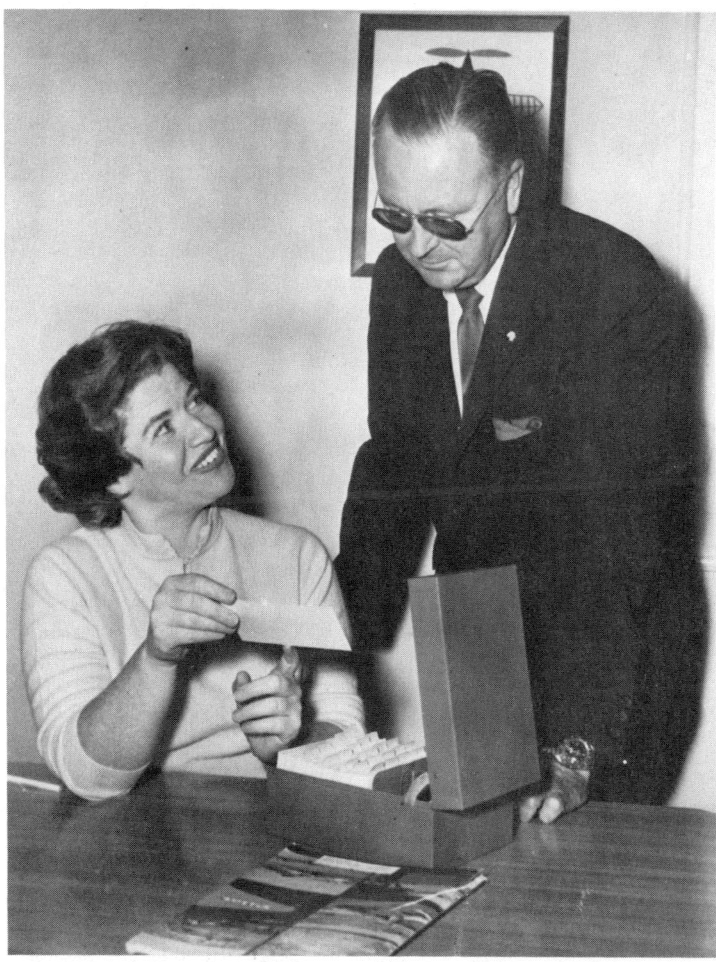

Fig. 5-3. Gil and Bob Gunnell, veteran Mexican group flight promotors, check a list of participants for one of their famed "Fly For Funsters" trips.

Fig. 5-4. A Mexican luau on the sands of a resort in Baja California is typical of the Gunnell "Fly For Funsters" trips. Everything is carefully arranged in advance. That's why Gunnell's groups have many repeaters.

using it ever since. Gil has raised three children in the process of putting trips together.

The shorter early trips took Gil a full working week, about $40 in phone calls and another $40 in mimeographing announcements and mailing. Most recent trips involve $150 or more in phone calls alone and up to $2000 just for music at the destination.

The pilots and their passengers who sign up for one of Gunnell's Mexican junkets can be assured of anything but boredom. There have been marriages on the trips, members almost jailed, earthquakes ridden out and, as Gil puts it delicately, "a number of marriages almost put in jeopardy."

Gil chuckles about the three women in one of her groups who admired a "gorgeous" Mexican male singer. A sort of wager was contrived among the trio as to who would get to know the Latin better. When it came time to depart two days later, one of the women was finally tracked down with the popular singer in or near his apartment. Her excuse for being late at the flight line was "she couldn't find her wig and she came here as a blonde and . . ." That's the Flying for Funsters during some of their less restrained moments.

Then there was a Halloween costume party in Mexico. One couple spent most of the day preparing headhunter costumes with ebony body paint. They missed the bus and tried to get a cab, but the driver called the police because he thought that the visitors were about to start a revolution.

Word came from the *calaboso* and the Gunnells talked patiently to free their painted Funsters.

On another occasion, a couple dressed up as Mexican farmworkers arrived late and local doormen wouldn't let them into the lavish banquet until Bob or Gil was called for verification. Needless to say, this couple won the costume competition.

During a visit to Guadalajara, Russian conductor Stowkowski conducted the local symphony orchestra at the famed old Delgollado Theater in one of his final concerts. On the spur of the moment, the Gunnells were able to reserve a block of seats for the fliers who wished to attend.

No matter how many times the Funsters return, Bob and Gil pre-visit the area some three months before the fly-in to make on-the-spot reservations and arrangements for music, boats, and buses, buffets and barbecues. Then, at least two days prior to the group flight, the Gunnells are back at the location to make doubly sure that everything is ready. On these planning trips, Bob and Gil try to get acquainted with local waiters, cab drivers, and bartenders. At San Carlos, near Guaymas, for example, they found a local little theater group performing. While it was nothing professional, they arranged for more than 100 Funsters to watch an English version of a traditional Mexican production.

During one fulldress banquet with more than 100 people at a Mexican hotel, everything was set and the marqueritas were flowing smoothly when Gil noticed her catered steaks and seafood being served in an adjoining banquet room to a group of flying physicians. She descended on the chef, pointed out his mistake, and replenished the bar until a new order of shrimp and ham could be prepared. Most of the guests didn't know the difference, and it developed into quite a party.

All Funster trips are by invitation only. "We don't ask anyone to go with us to Mexico until he has at least 250 hours and considerable time in the ship he'll be flying," said Gil. Newer pilots combine dual cross-country instruction with our pilots if they're short on experience. We also run a 'copilot's' course somewhat similar to the AOPA's 'Pinch Hitter' for apprehensive wives. One gal was so uncomfortable in a lightplane that we finally took her to a hypnotist, and then she did quite nicely. Many other graduates of our ten-hour course, taken in the family airplane, have gone on to get their license. One copilot graduate went ahead to get her multiengine rating and began flying the family 310."

Inexperienced copilots find a group therapy as Gil arranges informal meetings between veteran lightplane travelers and newcomers. "The gals get together and find out they're not the only ones who don't enjoy turbulence and the fears that go with it. They're able to talk with knowledgeable wives who have been there before and not rely solely on the comments of their own husbands. Some still hang on for dear life when it gets really choppy, but now they compare it with a free ride on a rollercoaster or a drive down a bumpy road and they survive," explained Gil.

Funster name tags are informal with doctors and company presidents tagged with only their names and no titles. There's a tiny silver charm bracelet given to each lady on each trip. Many veteran Funsters now carry a heavy bracelet with over 35 mementos of these trips. The charm bracelet idea dates back to the first Gunnell fun fly-in when ten planes and 35 people went to the Wigwam Country Club at Litchfield Park just west of Phoenix, Arizona. A local silversmith made up a tiny "wigwam" figurine. After three trips, men in the group are eligible for a set of lapel wings.

No matter how many times veteran visitors have watched, it's always a revelation to see Gil perform with enthusiasm and that intangible faculty to "wing it." Both veterans and first-time visitors comment that Gil missed her calling in not going in for vaudeville—or perhaps burlesque. She has a singular repertoire of jokes, some even printable, to fit almost every occasion. She can make an award for the closest to E.T.A. or the biggest flight booboo seem like an Academy Award presentation. She plans far in advance to have a presentation, ridiculous or serious, for virtually every planeload on every trip. She works from memory and a few notes; even at the end of a long day with the pitfalls of sunburn and open bars, she's good!

On the Zihuatanejo trip in November, 1975, the Funsters had an earthquake, just 50 years to the day after a former temblor had almost wiped out the Mexican town. However, none of the Funsters panicked, and fortified with another marguerita, the party continued.

The most extensive of Gunnell's productions was a two-week excursion through Mexico, Guatamala and El Salvador. Twenty-two airplanes carrying 46 people went along on this four-country junket. The trip began with a rendezvous at Mazatlán with a fiesta and a pilot briefing. Then the group made a 680-mile hop to Puerto Escondido, some 145 miles down the coast beyond Acapulco. Aircraft on this trip were half-and-half twin-engine, but five high performance Cessna 210s, a Bonanza, Commanche, 182, 177 RG and an Aero Commander 112A were included.

Next came a border-crossing trip to the Ilopongo International Airport in San Salvador with a tramway lift to the top of an ancient volcano where an imaginative coffee planter had transformed "Teleferico de San Jacinto" into a fascinating fairyland. After a day on the town, the Gunnell group flew back to La Aurora International Airport at Guatamala City for a frantic two-day tour before departing back into Mexico and its Yucatán Peninsula.

While the Funster's emphasis is on partying with liberal lacings of libation, the seasoned pilots involved have a quiet, respectful way of watching the eight-hour bottle-to-throttle rule. With a fairly long flight home on Sunday—or later for those who can stay over an extra day—Gunnell's guest pilots come out bright-eyed early on getaway day, though it might be mentioned in candor that some of the passengers carry scars of the night before.

Over these many years, the Gunnells have been able to develop their Funster trips into a flying package that is mostly social, partially travel with a pre-package wrapping that removes detail responsibility from both pilots and passengers—a unique product, and it's a very difficult act to follow (Fig. 5-5).

ANGEL DERBY TO ACAPULCO

As the popularity of flying in Mexico increases, many organizations are sponsoring group fly-ins. One such group flight is the Angel Derby, the All Women's International Air Race, established as a biennial competition sponsored by the Mexican National Tourist Council. A purse of $10,000 was established for the first competition in 1981.

The initial Mexican Angel Derby was set as a three-day affair with 45 single and twin-engine aircraft ranging from 145 to 580 horsepower competing for prizes and prestige. With a U.S. takeoff from Van Nuys, California, the women contestants were scheduled to cross the border at Mexicali and spend their first night at Guaymas. A second night's stop was scheduled

Fig. 5-5. Two members of a Gunnell flight admire the scenery at the Palmilla resort on the tip of Baja California before starting that 1000-mile trip back to Santa Monica, California, where these trips originate.

for Puerto Vallarta before completing the 1731-mile trip to Acapulco. Additional refueling stops were scheduled at Puerto Peñasco, Los Mochis, Mazatlán, Manzanillo, and Zihuatanejo for smaller aircraft with limited fuel range.

The first Acapulco Angel Derby was promoted in the tradition of Amelia Earhart, famous woman pilot who participated in the first women's air race 51 years ago.

Each of the aircraft entered in the race is handicapped according to type and size, according to the rules established by the Sporting Code of the Federation Aeronautique Internationale, which makes allowances for the different size and power of the aircraft entered in the race. All the airplanes entered must have non-supercharged engines of between 145 and 580 horsepower, but because of the handicap system, additional power does not provide an advantage, and the race is a test of pilot skill and ability.

The Miguel Aleman Trophies were scheduled to be presented to the winners at a special banquet in the Pierre Marques Hotel in Acapulco following the competition.

Chapter 6
East Coast

The east coast of Mexico, from Texas to the Yucatán Peninsula, really isn't all that long nor that glamorous. It's roughly 2½ times as far down the Pacific Coast to Guatamala as it is from Matamoros to Villahermosa on the Rio Grijalva where the Yucatán Peninsula more or less begins. The coastline is sandy with broad beaches and some interesting towns. However, until you get into the Yucatán Peninsula, the trip is mostly business with few resort areas.

The last time we flew the east coast of Mexico, we covered the entire country in a single short day with a Cessna 207. This was in the days just before the requirement for a landing at the first available International Airport. We departed Harlingen, Texas in N6401H, cleared Customs at Tampico, fueled both the airplane and ourselves at Minititlán, and stopped well before dark at Tapachula. Our total flight time was an even six hours, including backtracking forty miles at Minititlán to stay with the main highway that crosses the Isthmus of Tehuantepec.

When you consider that Brownsville is as far south as Los Mochis on the Gulf of California, or south of Loreto in Baja California, things fall into perspective.

Matamoros, (Fig. 6-1), one of Mexico's 15 largest cities, is a jumping-off place for many visiting pilots and it is a popular border shopping point for many tourists. All normal border crossing facilities are available for an initial landing at this airport.

The main highway to Mexico City swings inland almost immediately and heads for Ciudad Victoria. Pilots enroute to Mexico City or the Yucatán are apt to stay with the shoreline and follow the beach to Tampico, especially following the broad, uninhabited beach on the seaward side of the

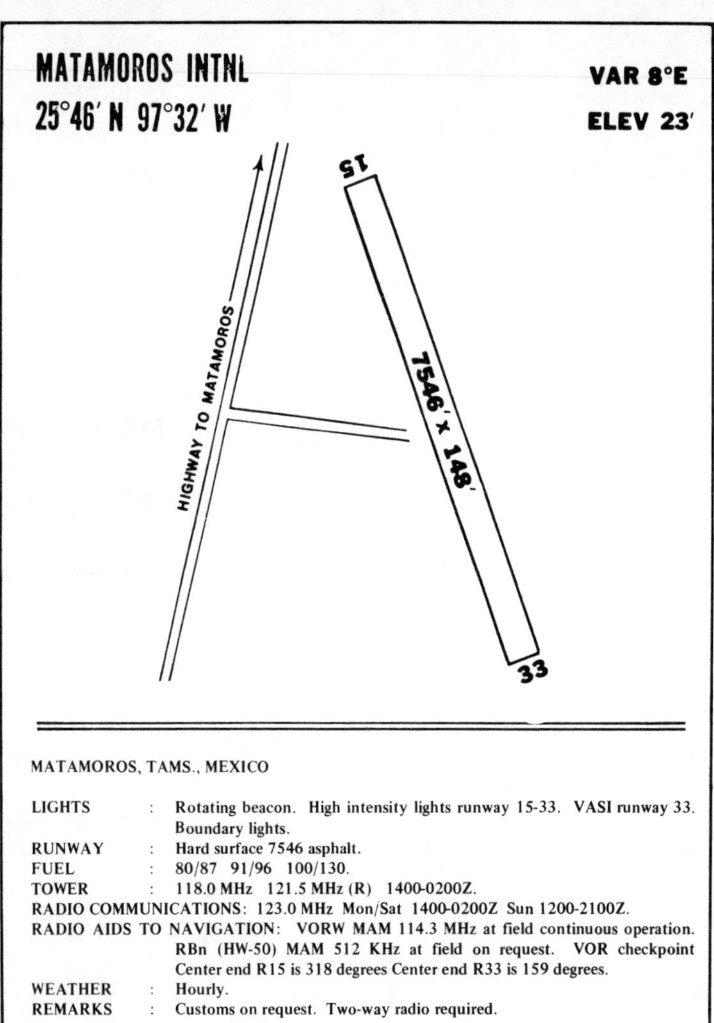

Fig. 6-1. Layout for the Matamoros International Airport. Many flights down the east coast will clear into Mexico here.

Laguna Madre (Fig. 6-2). We chose the shoreline for sightseeing at low altitude rather than climbing for optimum speed and fuel consumption on the relatively short 100 minutes to Tampico.

On the coast just east of Ciudad Victoria enroute to Tampico is a 2700-foot dirt strip called La Pesca. Inland a few miles is another dirt strip at Noche Buena on the Rio Soto la Marina. Both are operated by the Club Noche Buena, a fly-in fishing club. Check at Brownsville or Matamoros before departure for current information.

MONTERREY

While it isn't truly on the east coast, the area around Monterrey does not fit in the central portion of the country, so we'll mention it here. Pilots departing from Texas can best reach Monterrey from the border crossing airport of Nuevo Laredo across the Rio Grande from Laredo, Texas.

Monterrey (the second "r" was added to avoid confusion with the California town) is about 140 miles south of Laredo and has two large airports, both marked as "Airport of Entry" on the (1980) WAC CH-23. However, by the latest Mexican regulations, both of these airports are now listed as "interior airports" and should be used by general aviation aircraft only after clearing into Mexico at one of the eleven border airports. (Now the number has been increased to twelve, including Hermosillo, Sonora.)

Monterrey's General Mariano Escobedo International east of town is by far the larger of the two fields with a 9843-foot main runway (Fig. 6-3). Monterrey Del Norte, north of town, has a main runway 6600 feet long. The town is Mexico's leading industrial city with a population of nearing one million, so there may be sufficient air traffic to support both of these large airports. The usual pattern in Mexico, and elsewhere, is for airlines and corporate aircraft to congregate at the larger field and general aviation to stay at the smaller airport where not all the amenities are required. If you plan to visit Monterrey, check in advance on the relative status of these two airports (Fig. 6-4).

Industries in Monterrey produce glass, steel, furniture, elec-

Fig. 6-2. Beaches down the east coast of Mexico look similar to this scene where an Aztec C cruises the coastline.

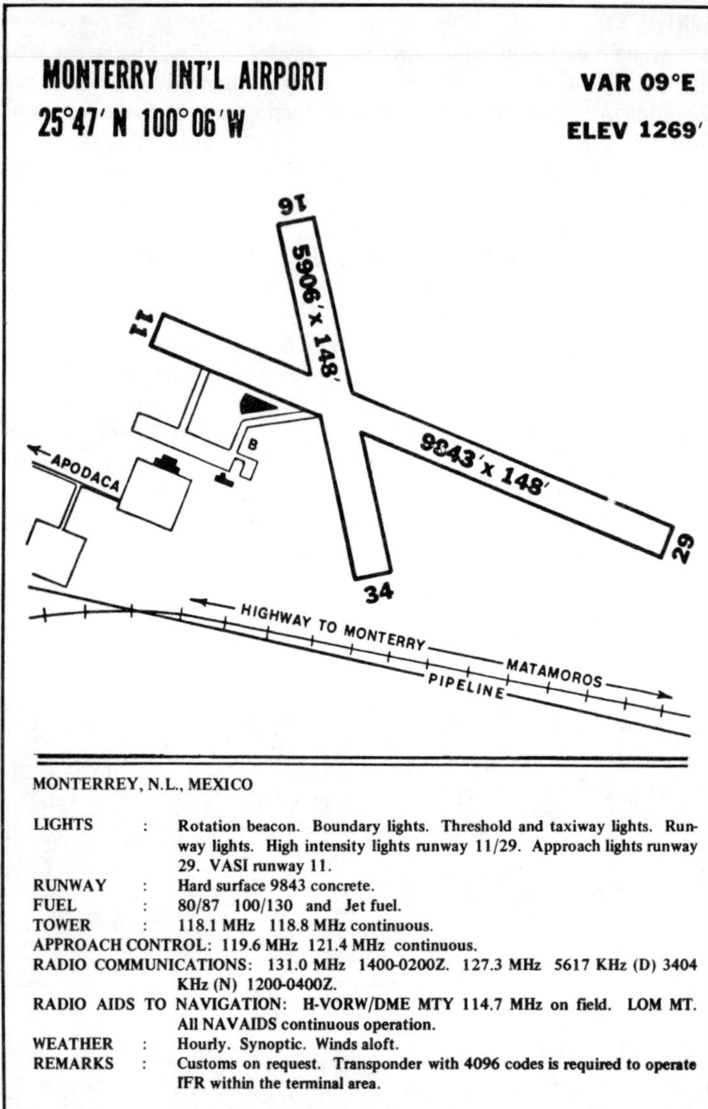

MONTERREY, N.L., MEXICO

LIGHTS : Rotation beacon. Boundary lights. Threshold and taxiway lights. Runway lights. High intensity lights runway 11/29. Approach lights runway 29. VASI runway 11.
RUNWAY : Hard surface 9843 concrete.
FUEL : 80/87 100/130 and Jet fuel.
TOWER : 118.1 MHz 118.8 MHz continuous.
APPROACH CONTROL: 119.6 MHz 121.4 MHz continuous.
RADIO COMMUNICATIONS: 131.0 MHz 1400-0200Z. 127.3 MHz 5617 KHz (D) 3404 KHz (N) 1200-0400Z.
RADIO AIDS TO NAVIGATION: H-VORW/DME MTY 114.7 MHz on field. LOM MT. All NAVAIDS continuous operation.
WEATHER : Hourly. Synoptic. Winds aloft.
REMARKS : Customs on request. Transponder with 4096 codes is required to operate IFR within the terminal area.

Fig. 6-3. Airport layout at Monterrey General Mariano Escobar International Airport. Check current status of this airport for International flights before planning to use it. Note on this drawing that Monterrey is spelled two different ways.

tronic equipment, and products of the Cuahtemoc Brewery, Bohemia and Carta Blanca, at 1,500,000 bottles per day. Sometimes called the "Pittsburgh of Mexico," Monterrey is also the capital of the state of Nuevo Leon and has two universities. Look for band concerts on Thursdays and

Sundays at the Plaza Zaragoza. The town has a number of large, expensive hotels, moderately-priced motels, and a wide variety of eating places. While car rentals are available at the airports, visitors should be cautioned that traffic is heavy, streets are narrow and there's very little downtown parking. You can, however, rent *calandrias*, little horsedrawn buggies for a tour of the downtown area.

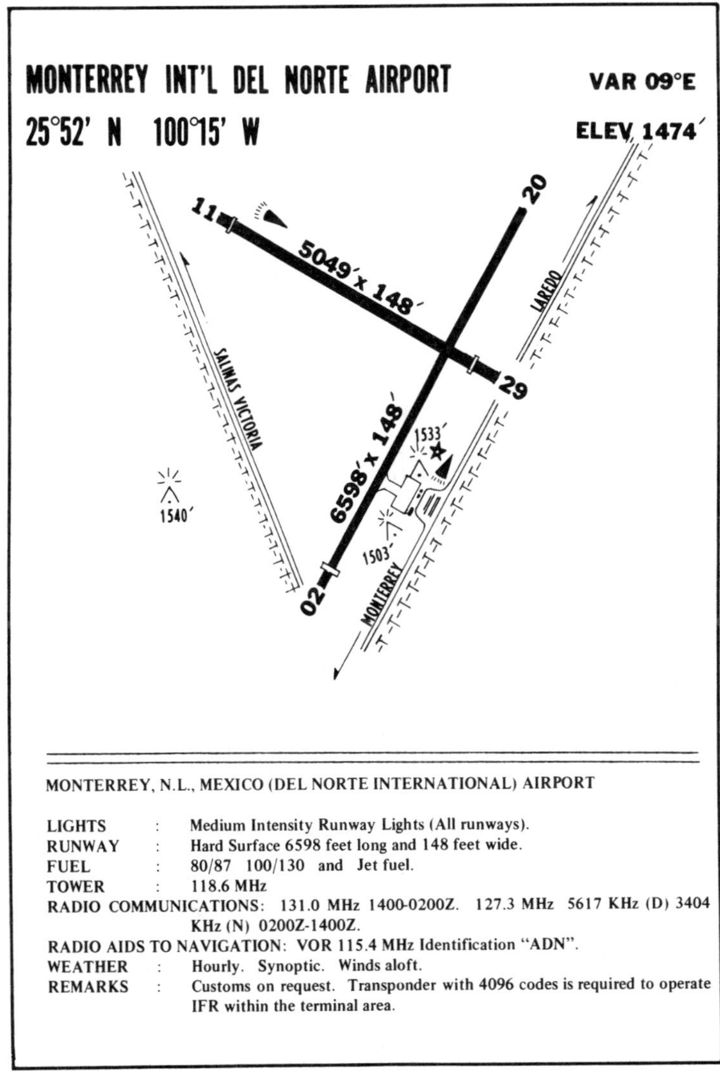

MONTERREY, N.L., MEXICO (DEL NORTE INTERNATIONAL) AIRPORT

LIGHTS : Medium Intensity Runway Lights (All runways).
RUNWAY : Hard Surface 6598 feet long and 148 feet wide.
FUEL : 80/87 100/130 and Jet fuel.
TOWER : 118.6 MHz
RADIO COMMUNICATIONS: 131.0 MHz 1400-0200Z. 127.3 MHz 5617 KHz (D) 3404 KHz (N) 0200Z-1400Z.
RADIO AIDS TO NAVIGATION: VOR 115.4 MHz Identification "ADN".
WEATHER : Hourly. Synoptic. Winds aloft.
REMARKS : Customs on request. Transponder with 4096 codes is required to operate IFR within the terminal area.

Fig. 6-4. Sketch of the Monterrey International Del Norte Airport. As with the other Monterrey Airport, check before flight for the current status of this as an International airport.

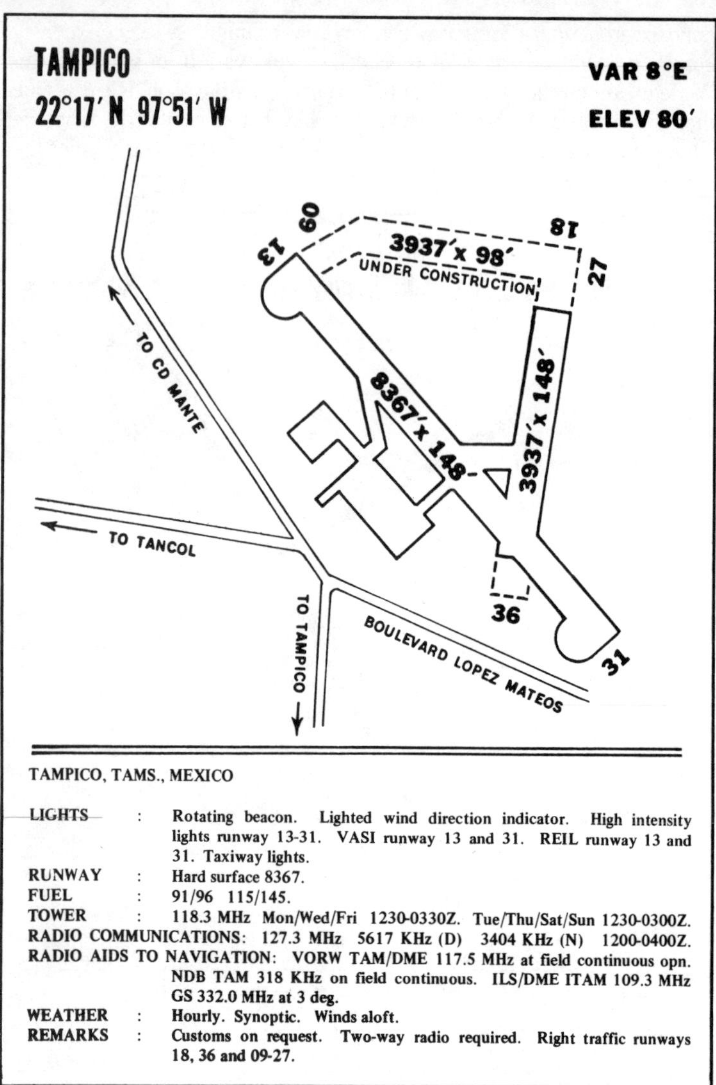

TAMPICO, TAMS., MEXICO

LIGHTS	:	Rotating beacon. Lighted wind direction indicator. High intensity lights runway 13-31. VASI runway 13 and 31. REIL runway 13 and 31. Taxiway lights.
RUNWAY	:	Hard surface 8367.
FUEL	:	91/96 115/145.
TOWER	:	118.3 MHz Mon/Wed/Fri 1230-0330Z. Tue/Thu/Sat/Sun 1230-0300Z.

RADIO COMMUNICATIONS: 127.3 MHz 5617 KHz (D) 3404 KHz (N) 1200-0400Z.
RADIO AIDS TO NAVIGATION: VORW TAM/DME 117.5 MHz at field continuous opn. NDB TAM 318 KHz on field continuous. ILS/DME ITAM 109.3 MHz GS 332.0 MHz at 3 deg.

WEATHER	:	Hourly. Synoptic. Winds aloft.
REMARKS	:	Customs on request. Two-way radio required. Right traffic runways 18, 36 and 09-27.

Fig. 6-5. Sketch of the Tampico airport shows an unusual layout of three runways. Surface winds in this area can be very strong and the triangular runway layout is designed to help landing pilots.

TAMPICO

The Gulf of Mexico can generate its own brand of weather. Tampico is well south of the Tropic of Cancer. In general, the hot, humid season begins in April and peaks in July, while storms and rain can be expected by September. There is a "Norther" season of strong, cold winds at the end of the year. In any event, get a good weather briefing before crossing the border and update as you can en route.

Tampico's International Airport and VOR are just north of town. The air carrier runway, 8367 feet long, is oriented on a heading of 31-13. Two shorter runways—both with right-hand patterns—make an into-the-wind landing possible most of the time (Fig. 6-5). Tampico is a good, dependable fuel stop for aircraft with limited range.

However, even tour books do not list Tampico as a tourist town (Fig. 6-6). Oil tanks and refineries line the Tampoon River. The AAA guide lists four motor hotels, two downtown and two in the suburbs. Best fishing in the area is for large tarpon in the rivers that drain into the Gulf. Hunters can find deer, quail, turkey and duck.

First ADF south of Tampico is Tuxpan with a paved 4600-foot airport The town lists 70,000 inhabitants and a major tarpon fishing tournament the end of each June. Just a bit farther down the coast and inland is the VOR at Poza Rica where the paved airport was marked "closed." This is Mexico's principal oil producing area. One well-known travel writer described Poza Rica as "a grimy oil-refining town with no visible attractions. If you must stop here overnight, the Hotel Poza Rica provided inexpensive rooms."

VERACRUZ

Your next logical stop would be Veracruz with two broad runways at the fine International Airport just east of town (Fig. 6-7). Veracruz is

Fig. 6-6. There is always shopping to be had, even in non-tourist towns like Tampico. This young lady bargains with a street merchant (courtesy National Tourist Council).

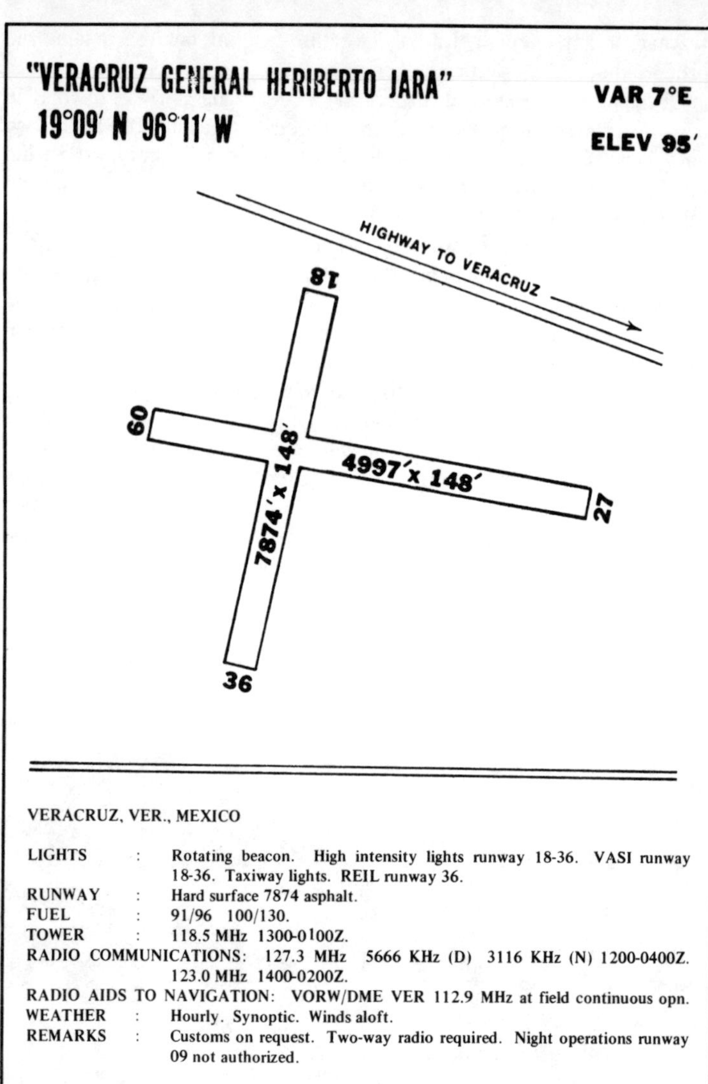

VERACRUZ, VER., MEXICO

LIGHTS : Rotating beacon. High intensity lights runway 18-36. VASI runway 18-36. Taxiway lights. REIL runway 36.
RUNWAY : Hard surface 7874 asphalt.
FUEL : 91/96 100/130.
TOWER : 118.5 MHz 1300-0100Z.
RADIO COMMUNICATIONS: 127.3 MHz 5666 KHz (D) 3116 KHz (N) 1200-0400Z. 123.0 MHz 1400-0200Z.
RADIO AIDS TO NAVIGATION: VORW/DME VER 112.9 MHz at field continuous opn.
WEATHER : Hourly. Synoptic. Winds aloft.
REMARKS : Customs on request. Two-way radio required. Night operations runway 09 not authorized.

Fig. 6-7. Sketch of the Veracruz Airport. The authors missed landing here because of a shortage of 100 octane fuel.

Mexico's oldest city and largest seaport. The area can be described as hot year-round and humid during the rainy summer season. If your timing is right, catch Carnival the week just before Lent where the town's Carib-Afro cultures blend in streets filled with dancers and musicians, costumed paraders, and overhead fireworks. This city of 300,000 has a selection of good hotels located both downtown and on the beach.

However, we were not destined to enjoy some of the great Veracruz seafood on our particular flight down the east coast enroute to South America. As we monitored the tower, we heard another U.S. aircraft being advised that the airport was out of 100 octane fuel, but this precious commodity was available at Minititlán. Then came a fast shuffle in the cockpit with plotters and other source material to find out what we could about our new destination. An hour and a half earlier there had been no information from the Tampico that this fuel problem existed when we filed for Veracruz.

But that's part of the unpredictable charm of lightplane flying. Somewhere along the trip, you're going to visit a place that you had never heard of before and perhaps really didn't want to see. Minititlán might just be one of those places.

We followed the main coastal road past Veracruz and listened to a steady stream of U.S. transient aircraft announcing their intentions to divert. We fervently hoped that Minititlán had an ample supply of av/gas. Weather began to deteriorate and we were soon flying under the bases of very turbulent, very wet clouds. Anything that wasn't tied down in the 207s spacious cabin began to float around a little and it was a time to figuratively roll up our sleeves and fly.

MINITITLÁN

Soon Minititlán's VOR began to show and we watched smoke from the oil refineries laying flat in a 30 mph-plus wind from the west. Our immediate problem was the single 5971-foot runway at Minititlán, which is built on a heading of 02-20, giving us a gusty crosswind that was higher than the proven component for our aircraft (Fig. 6-8). However, there wasn't any option, so we came in hot, with controls almost completely crossed to keep the big Cessna lined up with the runway and we deposited a little tire rubber in getting stopped. Soon a little Mooney duplicated our landing, followed by a Piper Aztec. The runway is 130 feet wide, but didn't look it.

We all congregated at the gas pit and were relieved to find that, even on a Sunday, the gas man was at the airport and fuel was soon flowing. Our CJ-25 WAC chart showed a new airport under construction some five miles northwest of the present in-town airport. Hopefully it will have a second runway for such windy conditions. While this town is conspicuously absent in surface-oriented tour books, we can assure you that, yes, there is a Minititlán, and yes, it has a good airport. There is a good-sized town with small hotels and good restaurants. Our Sunday lunch was in a family-oriented restaurant with good food well served on white tablecloths. During the season there is an enthusiastic crowd at the baseball game in the park just on the approach path to Runway 02.

Departure from Minititlán on a Sunday afternoon presents an unusual dilemma in lack of paperwork. The airport manager was not on the field and none of the gun-bearing mulitary officials waiting for a visiting air force general in a Lockheed JetStar were interested in preparing a flight plan for us. We finally asked the local AeroMexico Aeronaves manager to give us what weather information he had and to stamp our outbound flight plan. He provided the same service for the Mooney crew who were headed for Central America by way of Belize.

However, most pilots passing through the Minititlán-Villahermosa portion of the Isthmus of Tehuantepec would be headed for Yucatán, a flight covered in another chapter.

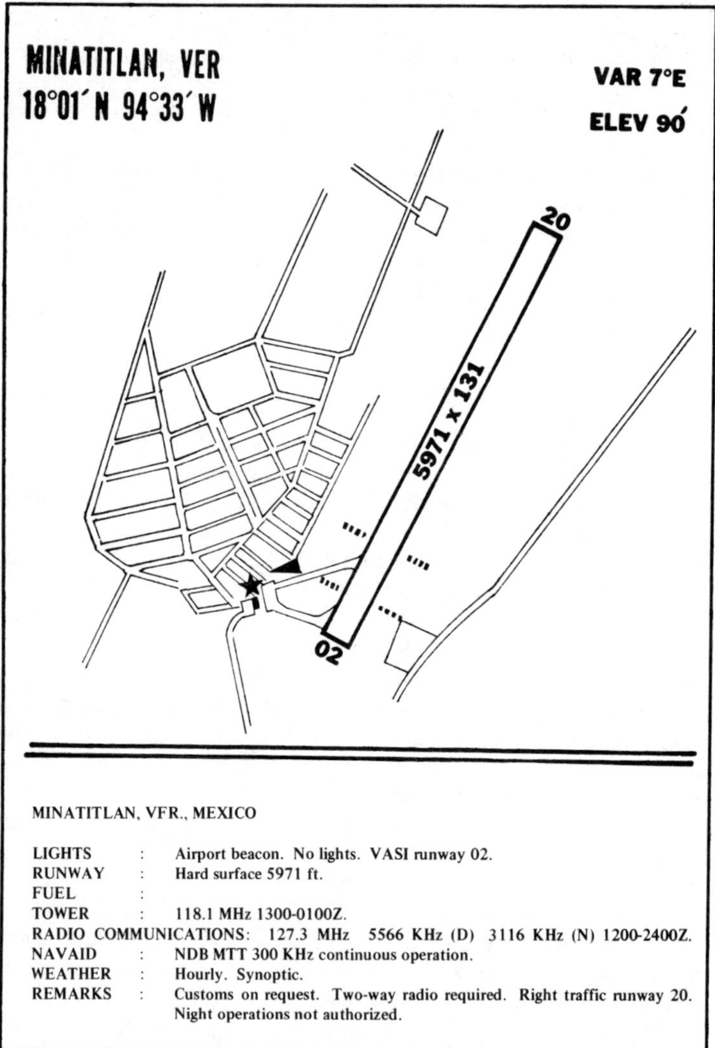

MINATITLAN, VFR., MEXICO

LIGHTS : Airport beacon. No lights. VASI runway 02.
RUNWAY : Hard surface 5971 ft.
FUEL :
TOWER : 118.1 MHz 1300-0100Z.
RADIO COMMUNICATIONS: 127.3 MHz 5566 KHz (D) 3116 KHz (N) 1200-2400Z.
NAVAID : NDB MTT 300 KHz continuous operation.
WEATHER : Hourly. Synoptic.
REMARKS : Customs on request. Two-way radio required. Right traffic runway 20. Night operations not authorized.

Fig. 6-8. Single runway layout at Minititlán can cause a problem in high winds.

ESCORT FOR THE RIDLEY TURTLES

A most unusual flying assignment in Mexico was undertaken by Gray Bower of Tampa, Florida. She and her daughter Jeanne spent two months working with the Florida Audubon Society at a conservationist field camp halfway between Tampico and the Texas border. She was the pilot and sole link between the field camp and "civilization" for a group assigned to count the arrival of the rare Ridley turtle (*Lepidochelys kempi*) that lay their eggs annually on a 15-mile stretch of beach near the village of Rancho Nuevo.

Here's a different type of flying in Mexico, as described by Gray Bower. She has been flying since 1974, has over 2700 hours with ratings through CFII, all ground ratings and is working on her Airline Transport Rating.

"I was called in several years ago to fill in as pilot for an American Bald Eagle Survey for the State of Florida. As the former owner of a J-3 Cub, I seldom had any reason to fly higher than 1000 feet and the 200 feet required for eagles did not faze me. I have been flying a Game and Fresh Water Fish Commission biologist on his annual flights ever since. Since low level environmental flying requires certain attitudes and skills, I have flown several types of assignments over the past three years through referrals and recommendations. I have flown for Florida Audubon as a fill-in pilot for their Manatee survey.

"Last year on a cold February day when I was scheduled to fly Eagles and Florida Audubon needed me for Manatees, I complained that they never offered me a real job that I could accept. So they asked about two months of roughing it in Mexico. It is lucky for me that my husband is an understanding fellow, for I'm not sure anything could have made me want to stay home instead. He did, however, engineer getting our 17-year-old daughter Jeanne out of school for the rest of the school year and sent her along as my chaperone on the premise that living together in a tent would give us a chance to get to know each other better (Fig. 6-9).

"I took over the responsibility of getting Florida Audubon's 1967 172 ready for two months of roughing it and Jeanne and I flew it to Brownsville. After a day of running errands and purchasing supplies, we packed up the aircraft and attempted to leave for Mexico. A terrible engine roughness convinced me that I was not going anywhere despite the Styrofoam chest of ice cream packed as a surprise for the Americans already at the camp at Rancho Nuevo. Hours later, after every mechanic in the house had given up, the test flight turned up nothing. We decided to depart the next morning so that I would have plenty of light when I landed on the dry lake bed behind the sand dunes at our beach campsite. It turned out that the only difference between a terrible roughness and no problem was a light misty rain. N3221L just loved to ice up in less than five seconds after starting in a light rain.

"We and the ice cream arrived at the camp the next day around noon and my landing strip had been pretty well cleared and marked. They had

Fig. 6-9. Gray Bower sits on the wing of her Cessna 172 as her daughter helps her refuel (courtesy Gray Bower).

even made me a windsock which I gratefully used for the next two months. We landed without incident and in the next few days I learned that it could be done on a regular basis. The 1200 feet of rough sandy soil required limited fuel, never more than two people aboard unless I had a good headwind, and the ability to tell when not to try it at all. No owner's manual tells you how to make an S-shaped 20° flap takeoff in 120° heat on a rough field. As for accelerate/stop distances, I regularly lifted off 20 feet from the end of the strip and just short of the lagoon. You had to know before you started that it could be done as by the time you realized you would not make it, you would not have the time or room to stop.

"We had an international crew at our camp. There were three American girls as research assistants, one American man as crew supervisor, five Mexican marines on 24-hour guard duty, several *Pesca* (Mexican Fisheries) biologists and inspectors, two graduate students from Holland, and groups of students from a University in Tampico. We were a fairly large and diverse group in a very isolated location, and we had to work hard at getting along.

"At first it was interesting to see who would be willing to go flying. The Americans were the first volunteers, then the most cosmopolitan Mexican project head from Mexico City. The next to fly with *la gringa* was the Marine Sergeant (macho, you know), but before the time came for me to leave, everyone had been up with me. They all loved it as I swooped over the camp and showed them the world from 200 feet. Even the workers from the village nearby went flying and, as I offered a ride as a birthday present to our cook, he could hardly refuse. Mexicans have a marvelous facility for mixing space-age technology with a lifestyle that has hardly changed in 2000 years. I enjoyed sharing my lifestyle with them as I shared theirs each day.

The duties of the aircraft were varied. N3221L and I were an emergency link with civilization, but mostly we attempted to track the seven Ridleys to which we had attached transmitters (Fig. 6-10). I had receiving antennas approximately 5′ × 3′ taped to the wing struts. While they affected my cruise, they seemed to lower my stall speed (indicated), and I regularly landed with 40 mph or less indicated. We were trying to learn about the behavior patterns of the Ridley in the vicinity of the nesting beaches. Our 15 miles of beach in a remote area north of Tampico is the only nesting site in the world for the Atlantic Ridley. The tracking was unsuccessful primarily because the Ridley did not surface for more than 3-5

Fig. 6-10. N3221L parked between tropical buildings near the Ridley Turtle research center in Mexico (courtesy Gray Bower).

Fig. 6-11. Gray Bower carries a boxload of turtle eggs to her 172 that she's just landed on this paved highway.

seconds which was too short a time for us to pinpoint location from the aircraft. We also used the aircraft to patrol the beaches for poachers. Unlike trucks and motortrikes which are very slow, the airplane could survey the nesting activity and dispatch the research crew and marines by walkie-talkie immediately. In addition to protecting the nesting turtles from poachers who would kill the turtle for the eggs (considered to be aphrodesiacs). We also tagged and measured the turtles. We marked all the nests which we then dug up at sundown and reburied in a guarded "corral" near our camp. We worked very hard to protect "our" Ridleys, and by the end of our assignment we felt we had contributed a great deal to a rare and vanishing species.

"One of my major assignments for the Ridley project was to transport some of the eggs to Texas. As the world's most endangered sea turtle, the Ridley is rapidly disappearing and the United States and Mexico have cooperated for several years in an attempt to preserve the species. The only nesting site in the world for this turtle, as I've said earlier, is the meager 15 miles of beach on the gulf coast above Tampico. Several attempts have been made to move some eggs north to the protected seashore of Padre Island in Texas. It is hoped that some turtles will someday return to nest on Padre Island where they are protected from human predators. A headstart project started in Texas by U.S. Fish and Wildlife hatches a few thousand eggs and then raises the hatchlings for a year or until they reach a size where their chances of survival are greatly increased.

"In an attempt to help them return to Texas, we caught the eggs as they were laid and before they touched the sand, then buried them in Styrofoam boxes of sand from Padre Island. When we collected enough, I was to fly the eggs to Texas. As an endangered species, many layers of government red tape had to be crossed before the paperwork was ready. We did not have a scale of any kind and each box varied somewhat, but we had 2000 eggs in 22 boxes of approximately 25 pounds each for a 500-600 pound load.

"My faithful 172 could never carry that much weight and fuel for a 200-mile trip off of my 1200-foot strip. So we hoped for some wind, left early in the morning while it was still cool carrying half a load and landed on a paved road which led to a fishing village about 15 miles south (Fig. 6-11). I hurriedly unloaded. I could not taxi off the side of the road because of the steep shoulders; the local bus was forced to pull off the road to go around us.

"I returned to camp for the rest of the eggs, landed on the road again and then loaded all the boxes into the aircraft. I had to climb over the boxes to get into the pilot seat and on takeoff it seemed very handy to have a runway that went on for miles. The takeoff was uneventful and I wagged my wings in farewell and I passed by the camp on my way north. Everyone was relieved to see me on my way and all I had left to worry about was the border crossing.

"Despite advice to simply overfly and land in Texas, I refused to ignore the new Mexican regulation which requires that you land in a border airport both coming and going. My Spanish has been improving but was still limited, so I carried a fist full of permits, a long list of names of the chiefs of this and that department with their telephone numbers, and an incredible sense of optimism. I arrived at Matamoros, followed by guide through the obligatory routine and was astonished when they said I could go. Not a word about the boxes piled head high in the airplane! Not one to argue with such luck, I headed for the 172, but was unknowingly followed by one final

Fig. 6-12. Fueling was done from the back of a four-wheel drive truck with 50-gallon barrels driven in from Tampico.

Fig. 6-13. During the various flights of this Cessna 172 in Mexico, field maintenance was required. Here a Mexican mechanic has pulled the cowling and is working on the Continental engine.

customs man who came to look. As I opened the door, he stared at the boxes and asked me, in Spanish, what was in them. '*Huevos*,' I answered. 'What kind?' he said. '*Tortugas marinas*,' I answered. As the horror of nearly letting me escape crossed his face, I hastened to assure him there was no problem and that I had all the papers. He sent for his supervisor. They looked at all those complicated papers and sent for their supervisor. Eventually I had about six of them trying to decide what to do with me. They finally decided I had too many official-looking papers and too many important names on my list to tangle with me, so they let me go—not, however, before I went back through the entire customs checkout routine, this time with all the i's dotted and t's crossed on their records.

As I crossed the border to Brownsville, I requested that U.S. Customs be notified to have a special agent from Fish and Wildlife standing by. It turned out that even the U.S. Customs Office in Brownsville had absolutely no desire to handle the paperwork for importing 2000 Ridley eggs and insisted that the special agent make sure that all the permits were in order. As I signed page after page, I got the impression that it should have been in blood. The eggs were successfully delivered to Corpus Christi where they were hatched and are currently growing to dinner plate size munching on turtle chow. They will be released this summer and there will be 2000 more young Ridleys with an excellent chance of survival.

"Fuel was always a problem for N3221L and me. We brought our fuel into the 1200′ strip in 50-gallon barrels from Tampico—a six-hour trip by four-wheel truck. I had a metal funnel with a car fuel filter in the bottom and we had a large hand-driven pump which we used to refuel the airplane (Fig.

6-12). The only time we had fuel contamination was the night it rained and we had not tipped the barrels so that water would run off.

"On one trip down the west coast we topped at Acapulco and headed south. Puerto Escondido did not have any fuel when we stopped for the night, and I did not believe that the dirt runway at Salina Cruz in southern Oaxaca had fuel the next day. I elected to call the tower at Ixtepec Military to ask but was greeted with silence. Too low on fuel to try both, I landed at Ixtepec and was met by the daily 25 knots of wind and the military police. Finally convinced we were simply stupid *gringos* (thank heavens for my flight plan from Puerto Escondido), they filled our fuel tanks from barrels, fined us, and sent us on our way. They did search the aircraft for guns and refused permission to take photos, but considering we really had no business there, we were very well treated.

"I never saw any contaminated fuel from airport fuel supplies and never paid any bribes, landing fees or even tiedown charges" (Fig. 6-13).

Chapter 7
Yucatán Peninsula

We first heard about the travels of Bernie Helgesen of Elgin, Illinois, from the pages of the Cessna 170 News. In late 1979, he penned a report about his solo visit to the tip of the Yucatán Peninsula, flying alone in his stock Cessna taildragger. Bernie's trip took place in June and July.

Details of Bernie's border crossing technique are covered in Chapter 3, so we'll pick him up departing Matamoros. Bernie made the two-week trip alone because "It's a long haul to Yucatán and I find it easier to fly solo than trying to make plans with someone because they don't always pan out." Bernie began flying in 1946 on the G.I. Bill and has owned five airplanes—three of them Cessna 170s. Here's how Bernie recalls his tour of the East Coast of Mexico (Figs. 7-1, 7-2).

MATAMOROS TO MÉRIDA

"From Matamoros I flew down to Ciudad Victoria where there's a VOR and a paved 4600-foot airport. There's great bass fishing around Lake Guerrero and I really had a ball. I'd read about this in one of the sports magazines, and it was one of my desires to try it. I only fished one day, but I caught about 20 in an hour; quit keeping track after that, but I must have caught at least 50 fish that day. I didn't get any of those great monsters that they talk about, but I did catch plenty of two-pounders. I kept maybe a dozen or so and I gave them to a Mexican boy who was hanging around the hotel when I came back in.

"I rode around mostly on my Honda 90. I carry it with me on all of my trips. I have it cut in two for storage in the airplane. It's built similar to a girl's bike where it has a single bar, and it's a very easy project to cut it and make a joint. Works real fine.

Fig. 7-1. The El Castillo Pyramid is part of the most famous of ancient Mayan cities located at Chichén Itzá on the Yucatán Peninsula (courtesy Mexican National Tourist Council).

"Down around Ciudad Victoria I met some people at the airport who were very friendly and one spoke English pretty well. We got to talking. I speak a little Spanish—I took a course up at the college before I left for conversational Spanish and it helped. I would recommend it, and it's kind of fun. Between my broken Spanish and their broken English, we had a lot of fun. They invited me to a real country-style *fiesta* for Mexican local people that actually work on the airport—about 40 of them—and I made many *amigos* that day. I still write back and forth. I sent them a bunch of pictures that I'd taken. I ate some things that I never thought I'd eat, but they were sure good. I was going to fly out that afternoon, but beer was served so I stayed another night.

"The next day I left Ciudad Victoria and flew to Tampico. It's only 120 miles, airline. I stayed overnight at the Posada Tampico Hotel, a beautiful place. It was close to the airport and had excellent food and beautiful grounds. If it wasn't an old Holiday Inn at one time, it sure reminded me of one. It had a double pool where you swam under a bridge into another pool—it was just a beautiful place. I stayed that night and flew the 260 statute miles to Veracruz the next day.

"Veracruz had plenty of gas—as did every place. I never had a bit of problem with gas; it ran about 60¢ a gallon [1979] and there was plenty of it. At Veracruz I met a fellow who happened to be working around an airplane. He was an American by the name of Jim Townsend and he was from out California way; he flew for one of the oil exploration companies. He came over and looked at my 170—he used to fly one. I asked him where to stay in town and he said come on with him. He had a little VW, so I didn't use my Honda that night. He took me to the hotel where he stayed and got me a room. There were about seven other fellows that he works with staying there. They operate the radar things on shore, and he flies grid patterns over the ocean and over the land exploring for oil.

"I stayed two days in Veracruz at that hotel right in town on the waterfront. The food was delicious and there was a good pool. Jim Townsend took me all over town and showed me around.

"After Veracruz, I went to Mérida (Fig. 7-3). I made one stop at Ciudad del Carmen, 320 miles, for gas and then flew another 210 miles to Mérida. The compass course from Ciudad del Carmen to Mérida is 040°, that is, northeast. Mérida was a nice clean city and I enjoyed it very much. Again there was a lovely motel with excellent food right there. Again I met some Americans—a fellow who had flown from Florida to Cancún and then up to Mérida. [The entire Yucatán Peninsula is on the CJ-25 WAC (World Aeronautical Chart) on a scale of 1:1,000,000.] Imagine that, he came across the Gulf in a 172. So here I was feeling like I had a lot of courage and I looked at this fellow and thought, Oh, brother!" We'll leave Bernie's account of his adventures at this point and pick him up later.

Pilots who have flown this Florida-Yucatán trip must land at one of Mexico's International Airports of Entry at Chetumal, Cozumel or Cancún

Fig. 7-2. Another view of the great El Castillo Pyramid which is topped by a temple to the feathered snake god Quetzalcotl. Each of the four stairways has 91 steps, adding up to the number of days in a year (courtesy Mexican National Tourist Council).

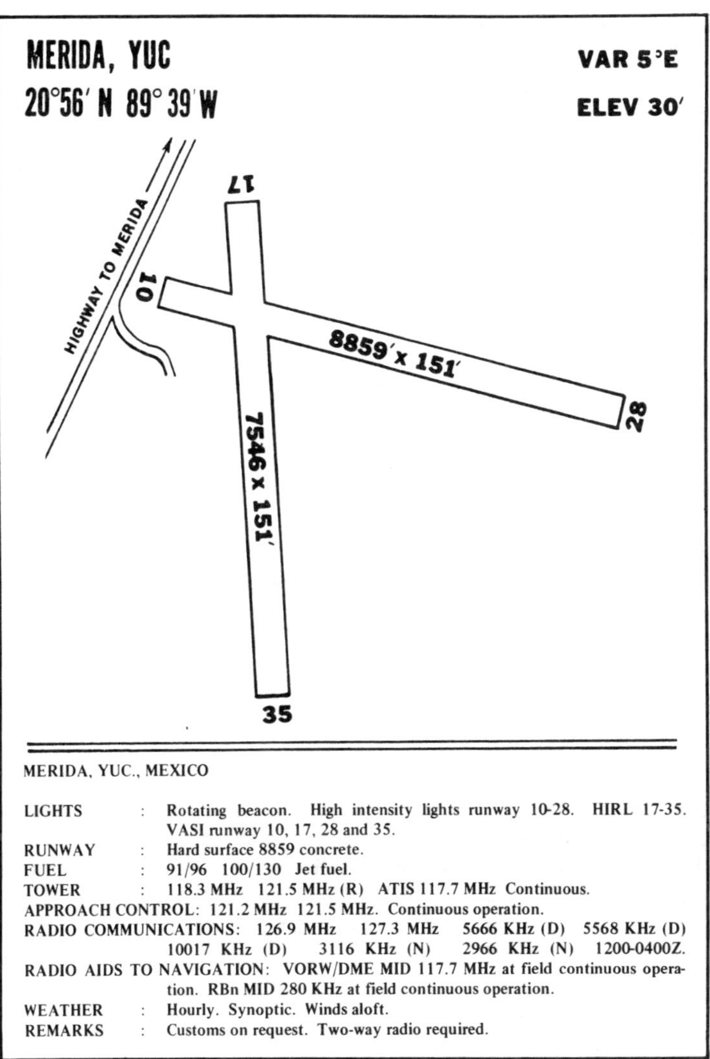

MERIDA, YUC., MEXICO

LIGHTS : Rotating beacon. High intensity lights runway 10-28. HIRL 17-35. VASI runway 10, 17, 28 and 35.
RUNWAY : Hard surface 8859 concrete.
FUEL : 91/96 100/130 Jet fuel.
TOWER : 118.3 MHz 121.5 MHz (R) ATIS 117.7 MHz Continuous.
APPROACH CONTROL: 121.2 MHz 121.5 MHz. Continuous operation.
RADIO COMMUNICATIONS: 126.9 MHz 127.3 MHz 5666 KHz (D) 5568 KHz (D) 10017 KHz (D) 3116 KHz (N) 2966 KHz (N) 1200-0400Z.
RADIO AIDS TO NAVIGATION: VORW/DME MID 117.7 MHz at field continuous operation. RBn MID 280 KHz at field continuous operation.
WEATHER : Hourly. Synoptic. Winds aloft.
REMARKS : Customs on request. Two-way radio required.

Fig. 7-3. Airport layout plan for Mérida on the Yucatán Peninsula. Mérida is the Capital of the State of Yucatán with 240,000 residents. It is the only town of any significant size on the entire peninsula.

(one and two piston-engine aircraft). With a twin, this routing would be ideal, and with the reliability of most of today's single-engine aircraft the overwater route is becoming more routine and less adventuresome.

From Key West on the tip of Florida, it's 850 nautical miles to the island of Cozumel. However, the route is offshore from Cuba and you work Havana Center enroute and are really never much more than 60 miles from land, even though a great part of that land belongs to Cuba. FAA regulations require an inflatable life raft and vests for this sort of overwater flight. This equipment can be rented at most large FBOs in the Florida area for about $15 per day.

Recent international visitors using this route report a $5 U.S. fee upon landing at Cozumel to spray the aircraft, $5 each for Customs and Immigration, and $5 to the Airport Manager for a "cruising permit." Bring your own tiedowns and check in advance with AOPA's Travel Department to find out about fuel. Cozumel is frequently out of 100 octane and Tulum, 35 miles distant, has no fuel at all.

Tulum, a 6100-foot paved strip near the shoreline, is unattended and ground transportation is reportedly not available. Pilots report no problem in walking to the ruins of the ancient walled city where licensed guides are available. Tulum is the only Mayan city known to be encircled by a wall and one of the few Mayan sites on the seacoast. *El Castillo* (the castle) is an outstanding structure and may have actually been a castle. The Temple of the Descending God has a carved diety over the main door descending head first. There are hundreds of little pyramids along the coastline north of Tulum.

The only hotel in Tulum is the Villa Arqueologica. Set in a jungle environment, it is close to many unexplored ruins. Electricity is furnished by a local hotel generator on a more-or-less regular basis.

CHICHÉN ITZÁ

We now return to Bernie's travels:

"After I left Mérida, I flew into a small 4100-foot paved strip by Chichén Itzá. The flight was only about 60 miles down a paved highway. The most famous ruins in Mexico are here, right in the heart of Yucatán (Fig. 7-4). This was the place that I really enjoyed. I stayed at a beautiful hotel right near the grounds; it's called the Mile End Hotel. You can walk to the ruins from this hotel, and you can walk to the ruins from the airstrip, although I didn't; I got my Honda out again.

"I highly recommend this stop. The ruins (I'm a history buff) were tremendous and I couldn't see enough of them. I stayed there two days climbing around them, in and out, and shooting pictures and what have you.

"One Mexican Tourist Bureau press release points out that the most popular of the ruins of the Mayan civilization are at Chichén Itzá. First founded in 432 AD, it shows the influence of the 10th century invasion by the Toltec Indians from central Mexico. Intricate plumed serpent designs honor the Toltec man-god, Kukulcan. The city was abandoned in the year 608, refounded in about 960, prospering until about 1200, when the Aztecs came upon it.

"Chichén Itzá was partially uncovered by a Carnegie Tech expedition over a twenty-year period. There are hundreds of buildings at the site and,

Fig. 7-4. Intricate detail is shown in ancient Mayan ruins throughout the Yucatán Peninsula. The area was developed in modern times by the growth of henequén, a form of sisal used in making rope (courtesy Mexican National Tourist Council).

at present, about thirty have been uncovered enough to be worth visiting. They are a breathtaking peek into the grandeur of the Mayan past (Figs. 7-5, 7-6).

"While unattended for all practical purposes, the airport at Chichén Itzá is watched over by the Hotel Mayaland, a luxury resort with good to very good hotel rooms, most with private balconies overlooking the ruins, two cocktail bars, swimming pool, shops and after-dark entertainment. Rates for this posh hotel in the archeological zone are higher than the Hacienda Chichén, also in the archeological zone, the Club Med's Villa Arqueologica, five minutes from the zone, or the Mission Chichén-Itzá, a mile from the zone.

"Chichén-Itzá is the only place where I didn't file a flight plan because there's no place to do it. When I filed a flight plan from Mérida to Cancún, I put on it "with a stop at Chichén Itzá." Well, I was there a couple of days and it worked out all right. When I ended up in Cancún, another 100 miles over a paved highway, I just told the Comandante that I had come from Chichén Itzá and he said, "That's okay. I know you can't file a flight plan out of there. That's the proper way to do it." And there was nothing more said about it."

CANCÚN

"At Cancún there's an 8500-foot international airport (Fig. 7-7). Cancún is a brand new city and it has everything—beautiful hotels, plush

Fig. 7-5. One of the complex ruins located at Chichén Itzá. Reportedly this Mayan civilization was founded in 432 A.D. (courtesy Mexican National Tourist Council).

private beaches. I shouldn't say *private* beaches because one thing I can say about Mexico is that all the beaches are open. It's none of this get out of here, get of there stuff. If you've got access to the beach, you can go to it. I've gone up to the hotel on my motorcycle and they usually have a place to eat and sit on the beach, and I've enjoyed it as if I was staying there.

"I stayed at a hotel in town which cost about $20 a night. I went to the beach every day and enjoyed beautiful scuba diving, snorkeling and swimming. It's just like the tropical scenes you see in the movies with the palm trees and beaches. I even climbed up and got me a coconut—there sure are plenty of them. The natives kind of laughed at me going up there; one felt sorry for me and threw me down a couple.

"I made more friends and I even went parasailing. You get towed

Fig. 7-6. Tourists descend the 91 steps of El Castillo Pyramid at Chichén Itzá. Note detailed masonry in this ancient ruin (courtesy Mexican National Tourist Council).

behind a boat in a parachute. They take you out over the ocean; you get about a 15-minute ride, I guess for $10. It's quite a thrill! I saw an old lady do it and I figured if she could do it, I could. I recommend it for everybody. You take off right there on the sand and you're flying as soon as they start you. When you come back, they slow the boat down and you land right where you took off from. It's very easy."

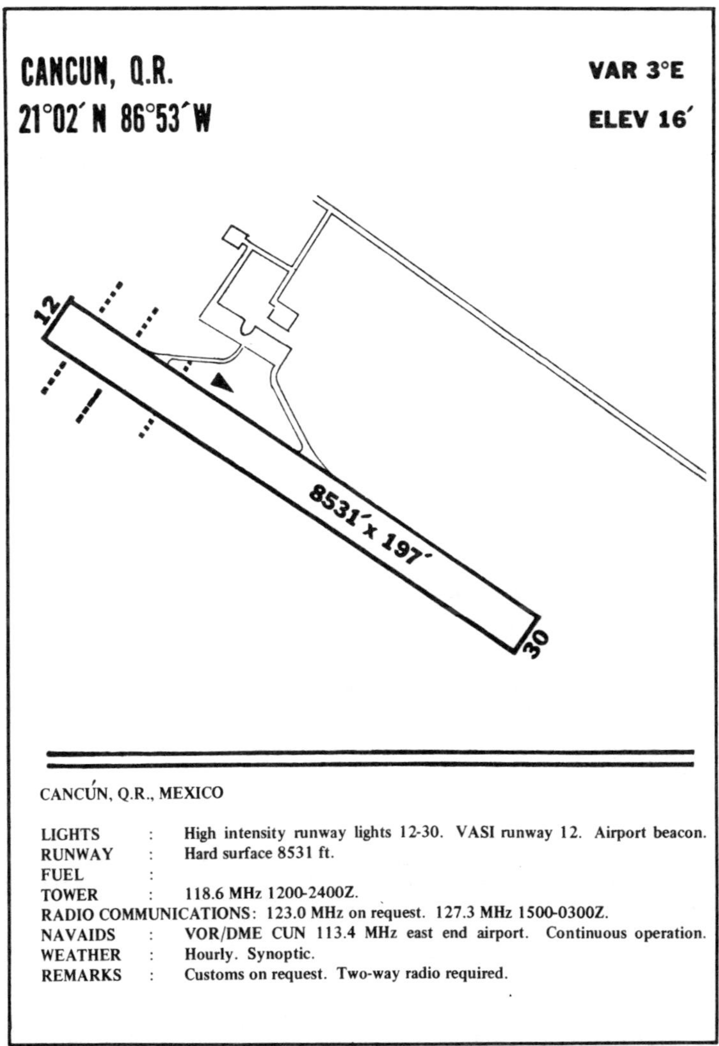

Fig. 7-7. Airport layout at Cancún, Mexico's newest major resort on the east coast. Cancún is an island 12 miles long and 500 yards wide in the Caribbean about 50 miles north of Cozumel.

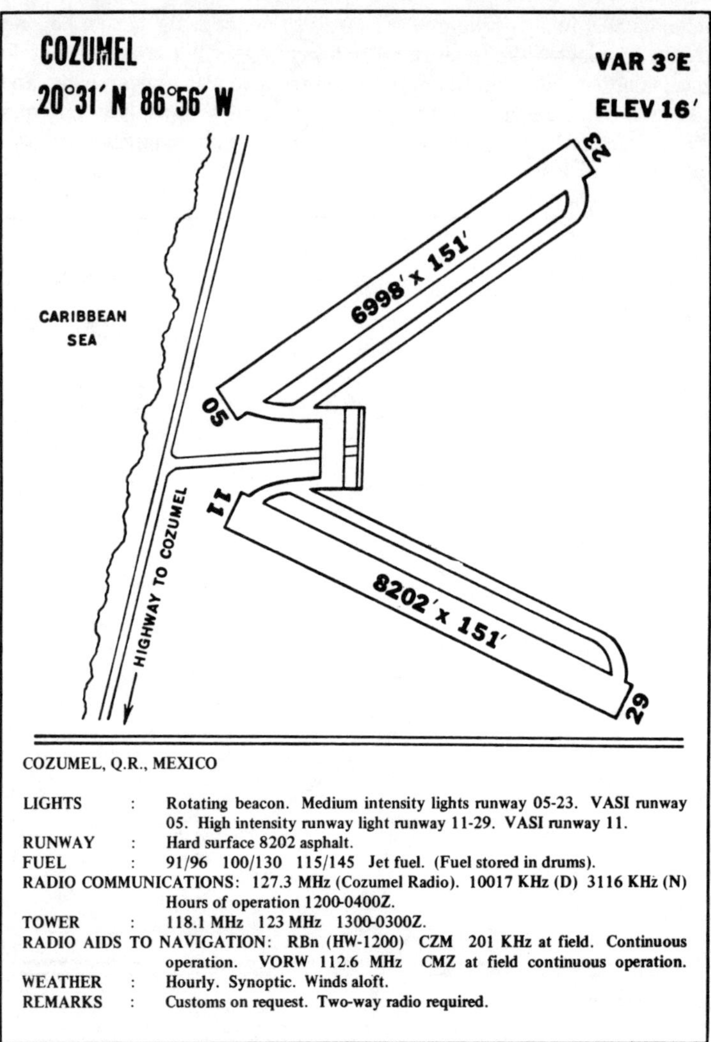

COZUMEL, Q.R., MEXICO

LIGHTS : Rotating beacon. Medium intensity lights runway 05-23. VASI runway 05. High intensity runway light runway 11-29. VASI runway 11.
RUNWAY : Hard surface 8202 asphalt.
FUEL : 91/96 100/130 115/145 Jet fuel. (Fuel stored in drums).
RADIO COMMUNICATIONS: 127.3 MHz (Cozumel Radio). 10017 KHz (D) 3116 KHz (N) Hours of operation 1200-0400Z.
TOWER : 118.1 MHz 123 MHz 1300-0300Z.
RADIO AIDS TO NAVIGATION: RBn (HW-1200) CZM 201 KHz at field. Continuous operation. VORW 112.6 MHz CMZ at field continuous operation.
WEATHER : Hourly. Synoptic. Winds aloft.
REMARKS : Customs on request. Two-way radio required.

Fig. 7-8. Airport layout at Cozumel. This area is very popular with diving enthusiasts with Palancar Reef located offshore, the second largest reef in the world.

COZUMEL

"After Cancún, I flew a little further out to an island called Cozumel with an 8200-foot international airport (Fig. 7-8). It's less than 40 miles and about 10 miles off shore. This was worth going as far as I did just to see this place. There's a hotel row on the Island of Cozumel with five very posh establishments: the government-run El Presidente, El Cozumeleno,

Cozumel-Caribe, Cabañas del Caribe and Playa Azul. All have air conditioning, private beaches, restaurants and bars. An additional five hotels are located in town with rates about 25% that of the larger hostelries. Again, I stayed in town. The hotels at the beach were very plush, but in town was where everything was at. A small hotel with a good restaurant was $15, right on the waterfront. I want to go back (Fig. 7-9).

"There was a Robinson Crusoe cruise for about $10 a day where they take you out on a boat in the morning in a small group and go around the island and find a small deserted spot. There you go snorkeling or scuba diving for your lunch. Lobsters or fish or whatever you get becomes the menu. If you don't get it, the crew will go get it for you. One guy found a beautiful starfish; it must have been a foot across. I wish I could have found one like that myself. When you get your fish, you go on the beach. They build a fire and you cook it. They bring everything else—the drinks and other food—so you get plenty to eat. It really was a fun thing to do.

"Another thing to do at Cozumel is to take a ferry boat out to another island called Isla Mujeres just off shore. There is also a 4100-foot paved airport there. Isla Mujeres is only five miles long and one-half mile wide. We went out to a place called Garafon Beach Reef which had excellent snorkeling and scuba diving. You're not allowed to keep any of the fish because they keep it for a scenic spot, but you can eat out there. It's just a beautiful setting. The reef is close to the beach and just loaded with all kinds of fish.

"Located 100 yards directly out from the La Ceiba Hotel Pier in Cozumel is the wreck of a 40-passenger Convair airliner lying upside down on a white sand bottom. The wreck is situated at the north end of the

Fig. 7-9. Street scene in Cozumel, reported to be the most unspoiled resort area in Yucatán despite jet flights from Miami and Mexico City. Hotels along this street in town cost as little as 25% of the fee at the new high-rise establishments. Hotel El Marques, shown here, is not even listed in the latest AAA Guide (courtesy Bernie Helgesen).

Fig. 7-10. Intricate stone carvings blend into jungle on the Yucatán Peninsula. Visitor Helgesen reports that nothing was removed from his parked aircraft. Would you think that these foreboding statues were a silent part of the local security force? (Courtesy Mexican National Tourist Council.)

Paraiso Reef formations. The plane was purposely sunk in June 1977 as a prop for a Mexican disaster movie entitled *Survive II*. Since then the wreck has become a popular dive site for both scuba divers and snorkelers. It is ideal for underwater photography because of the extremely clear water. Although the plane's engines have been removed, the rest of the wreck remains substantially intact.

"I used my Honda to go everywhere. I would put a lock and chain on it at night in front of the hotel on the street. No one even gave it a second look because you see hundreds of them around. I've seen as many as four people riding on one—father and mother and two kids. They go out on dates with them; you can rent them. They have them all over Mexico.

"The airports at both Cancún and Cozumel were excellent facilities. Everything is fenced; they keep everybody out. I didn't feel any qualms about leaving anything. My plane was full of stuff all the time. I didn't leave the cameras and stuff in plain view. I did have camping equipment. Nobody bothered anything (Fig. 7-10).

"With regard to weather information, they don't have the reports like we do here where you can look at them. The Comandante will read it to you. Nothing elaborate; the Comandante says, 'It's clear,' or ' its raining.' It's either good or bad and that's it!

"If you depart Cozumel for a flight back to Florida, seasoned pilots will call Miami FSS long distance (about $10 U.S.) for a weather briefing since there's no FSS or weather station at the airport. Aero Mexico's operations desk has a teletype with current sequence reports if you can read them. The local airline dispatcher is reported to be most cooperative with whatever meteorological data he has available. Then you open your international flight plan with Cozumel Radio and lots of luck. Personally, that's too much water for me."

Bernie went on to Belize (British Honduras), checked into the country and went on to San Pedro Island which he describes as "small, quaint and nice with good fishing and snorkeling." The Belize International Airport has a TCA marked on the charts. San Pedro has a 2600-foot dirt strip marked on the charts at an elevation of two feet. Great Britain has agreed to grant Belize independence circa 1980 as soon as jurisdictional disputes with Guatamala have been settled.

Returning to Mexico, Bernie flew back across the Yucatán jungles to Veracruz. He says that "anyone with an ADF can do it. Even without ADF, there's no problem because there are both roads and railroad tracks. They take you out of the way, but you can follow them. There are many ranches in this area, so it isn't too lonesome. The ADF is strong down there, even the local stations."

Note: This route is fairly close to the Guatamalan border. The WAC chart carries this statement: "Warning. Do not enter Guatamalan territory without prior radio contact with Guatamala Radio."

From Veracruz, Bernie headed back up the coastline to Matamoros, Brownsville, and on in to San Antonio, Texas, before calling it a day. "It's a long haul in a Cessna 170, but I enjoy long hauls."

Bernie Helgesen noted that there is not a lot of lightplane flying in the Yucatán Peninsula. "Most of the planes down there are working for a living. A small percentage of the people spoke English. I had a phrase book with a little Spanish and I got along just fine. The ground facilities, particularly at Mérida, Cancún and Cozumel are excellent."

Helgesen's entire trip continued as far west as Fresno, California, to attend an annual 170 Convention and on to San Francisco to visit his son. From there he returned to Illinois. He took a month and flew 94 hours to cover 10,000 miles. His report in the *170 News* concluded: "N170LB performed great—never skipped a beat. I would say that if anyone is planning a trip to Mexico, go; you will have a great time. The people are very helpful, kind, and courteous. Prices are very reasonable, food is excellent, and gasoline is plentiful at 60¢ per gallon for 80 octane (in 1979)."

OIL EXPLORATION FREIGHTERS

Somewhat like the back country in Alaska, the remote areas of Mexico are proving that air transportation is just about the only way to go. If you're planning to use one of the dirt strips in the oil-rich area of Yucatán, for example, you might keep an eye to the sky for the PEMEX "Super Herk," the huge four-engine Lockheed turboprop freighter that hauls oilfield equipment from Houston directly to where it is used in Mexico (Fig. 7-11).

Fig. 7-11. Aerial freighters like this PEMEX super Hercules turboprop transport frequently fly oil well supplies directly to the isolated areas of Yucatán (courtesy Lockheed Aircraft Corp.).

According to estimates, the reserves in Mexican oil fields stand at more than 60 billion barrels with a potential of up to 250 billion barrels. Large cargo transports like the "Super Herk" have been used in Mexican oil fields since 1977 when PEMEX contracted with Southern Air Transport (SAT) to airlift oilfield supplies directly to airports at Villa Hermosa, Ciudad del Carmen, and others in Yucatán. When the oil well first "blew out" in the Gulf of California, it was a SAT Hercules that hauled in heavy equipment to enable famed oilfield trouble-shooter "Red" Adair to stop the spill. Emergency cargo, including drill pipe, dispersent, machinery, tires, drilling mud, and cement is regularly loaded from ships in the Houston harbor and delivered directly to the oil fields of Mexico by Hercules.

Lightplane pilots operating in the vicinity of these heavy transports with up to 100,000 pounds of cargo aboard should remember that a very strong wake vortex can be expected. A landing behind such an aircraft should be at least five miles in trail and preferably above the flight path of the freighter.

Chapter 8
Central Area

The route from El Paso, Texas, to Mexico City is an interesting flight not normally planned by visitors. Either coastline is a simpler flight and more picturesque, but flying down the backbone of Mexico is a little more challenging than either low-altitude-coastline route.

The XIX Olympic Games were held in Mexico City in October, 1968. Shortly before the event was scheduled, we were able to "promote" a then-brand-new Cessna Skylane, N3109R, to fly the route to Mexico City and report on the problems involved. As any other flight into Mexico, it was an interesting experience.

We picked up the new Skylane from Cessna's distributor in San Francisco and ferried it to El Paso, Texas. After a careful briefing by the FAA FSS, we could have made the ten-minute hop (now required) and landed at Ciudad Juarez (Fig. 8-1), the closest airport of entry. However, at the time this flight was made, Ciudad Chihuahua was an international airport and we went that far on the first leg of this international flight (Fig. 8-2).

Even then, the tower spoke good English and we landed to find that Customs, Immigration, and Health were operating on a strictly 9-5 operation. So we went to town for the night and did the paperwork the next day.

CHIHUAHUA

Chihuahua is both a charming old city and state capital. Its economy comes from the rich silver mining, timber, and ranching in the area. *Chihuahuitas*, the very small dogs (Chihuahuas) are bred here. If you wish a change of pace, you can board the Chihuahua-Pacific Railway and travel across the rugged Copper Canyon country to Los Mochis near the Pacific

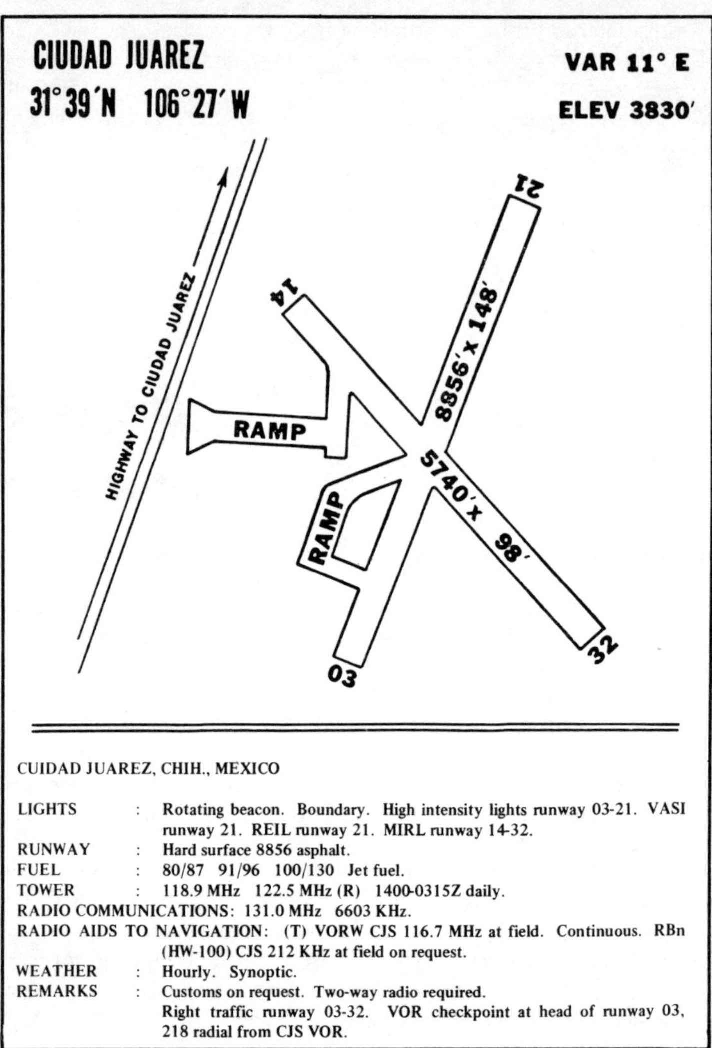

CUIDAD JUAREZ, CHIH., MEXICO

LIGHTS : Rotating beacon. Boundary. High intensity lights runway 03-21. VASI runway 21. REIL runway 21. MIRL runway 14-32.
RUNWAY : Hard surface 8856 asphalt.
FUEL : 80/87 91/96 100/130 Jet fuel.
TOWER : 118.9 MHz 122.5 MHz (R) 1400-0315Z daily.
RADIO COMMUNICATIONS: 131.0 MHz 6603 KHz.
RADIO AIDS TO NAVIGATION: (T) VORW CJS 116.7 MHz at field. Continuous. RBn (HW-100) CJS 212 KHz at field on request.
WEATHER : Hourly. Synoptic.
REMARKS : Customs on request. Two-way radio required.
Right traffic runway 03-32. VOR checkpoint at head of runway 03, 218 radial from CJS VOR.

Fig. 8-1. Diagram of the Ciudad Juarez International Airport just across the border from El Paso. Pilots entering here for the first time will probably find the time well spent to stop at El Paso FAA FSS before crossing the border to receive an update on rules, regulations, and fuel availability. The Texas State Aeronautics Commission puts out an excellent briefing book on Mexican flights.

Coast. The trip is picturesque and well worth the time to traverse it one way. We've taken this "Chihuahua choo-choo" and recommend it (Fig. 8-3). Recently, one or two air taxi companies flying light twins have been offering a flight between these two towns, making it possible to take the 14 to 18-hour train trip, most of it in daylight, and fly back in an hour or so.

An interesting tourist attraction in downtown Chihuahua is Quinta Luz, the home of Pancho Villa. It was a 50-room mansion but has been converted into a museum. The building has secret passages from the center to various exits and the bullet-ridden car in which Villa was ambushed and killed is also on display.

It was after noon before we were again back in the air on the 223-km hop to Torreon (Fig. 8-4). One of the main advantages of this route is that it follows the main highway from El Paso to Chihuahua, Torreon, Aguas

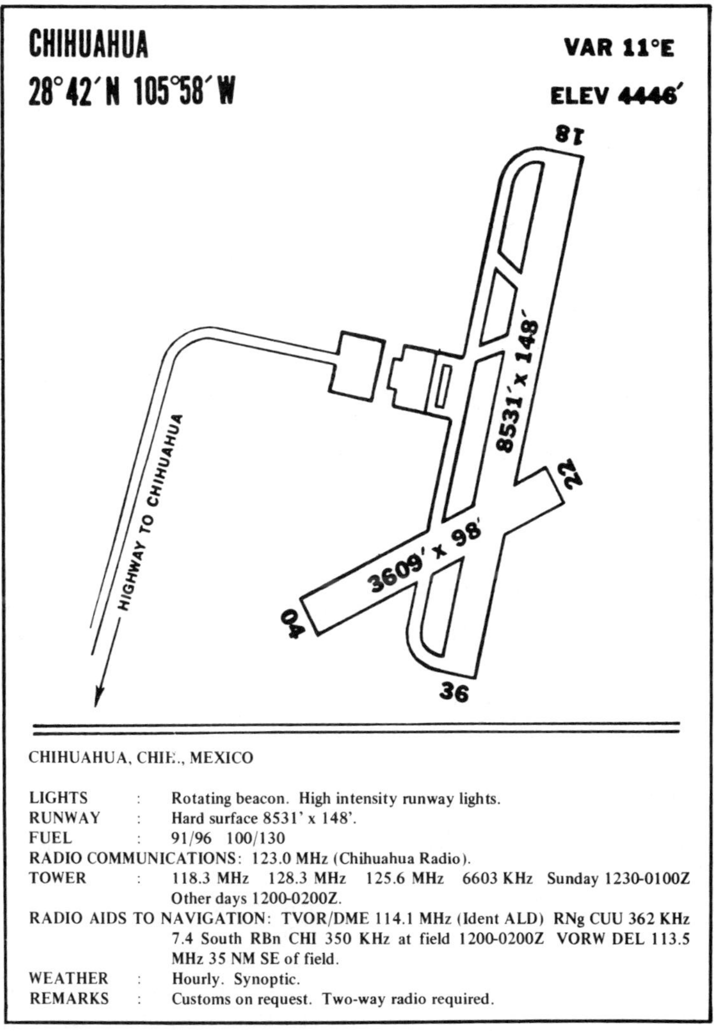

CHIHUAHUA, CHIH., MEXICO

LIGHTS	:	Rotating beacon. High intensity runway lights.
RUNWAY	:	Hard surface 8531' x 148'.
FUEL	:	91/96 100/130
RADIO COMMUNICATIONS:		123.0 MHz (Chihuahua Radio).
TOWER	:	118.3 MHz 128.3 MHz 125.6 MHz 6603 KHz Sunday 1230-0100Z Other days 1200-0200Z.
RADIO AIDS TO NAVIGATION:		TVOR/DME 114.1 MHz (Ident ALD) RNg CUU 362 KHz 7.4 South RBn CHI 350 KHz at field 1200-0200Z VORW DEL 113.5 MHz 35 NM SE of field.
WEATHER	:	Hourly. Synoptic.
REMARKS	:	Customs on request. Two-way radio required.

Fig. 8-2. Sketch of the Chihuahua City Airport. All ASA airports in Mexico are large enough to handle DC-9's.

Fig. 8-3. Chihuahua-Pacific railroad trip is a unique travel experience. It may be difficult to schedule for visiting pilots, but try this tour if you can (courtesy Mexican National Tourist Council).

Calientes, Leon, Queterero, and on into Mexico City. On this middle-of-the-century flight, the hop over the high mountains into the Mexico City area is very short.

On this particular flight, we spent exactly 48 minutes on the ground at Torreon for fuel, flight plan, landing fee (in those days there was one), taxi out and get back into the air. We detoured to the high-altitude, 6200-foot airport at Durango for an interesting evening in this vigorous silver mining area of Mexico. Durango has grown since this visit and at last report has a population of over 200,000. It is a state capital, now used by a number of Hollywood studios as a location site (Fig. 8-5). Durango has a cathedral dating back to 1695 that was completed in 1750 where bell ringers in the towers are visible from the plaza. Durango is off the beaten path or airway and well worth a visit. The jet-age airport here is at an elevation of 6093 feet, lighted, and now 8800 feet long. The old airport that we used on the Olympics visit was on the southwest side of town and has since been closed.

Next stop was San Blas, up over the Sierra Madre to 12,000 feet to top the hills. We let down leisurely south of Mazatlán and landed at what was then the smooth grass strip at San Blas. At that time there was a Mexican guard with a shiny new bicycle who would watch over your airplane for 25 pesos per night (Fig. 8-6). He was raising five children on the fees collected from aircraft watching. Upon our return after two days of sightseeing, we saw as the only encroachment a set of muddy pussycat tracks over one stabilizer.

From San Blas, we circled just over the hill and fueled at Tepic. Here our three-day-old flight plan was closed with no question since San Blas had

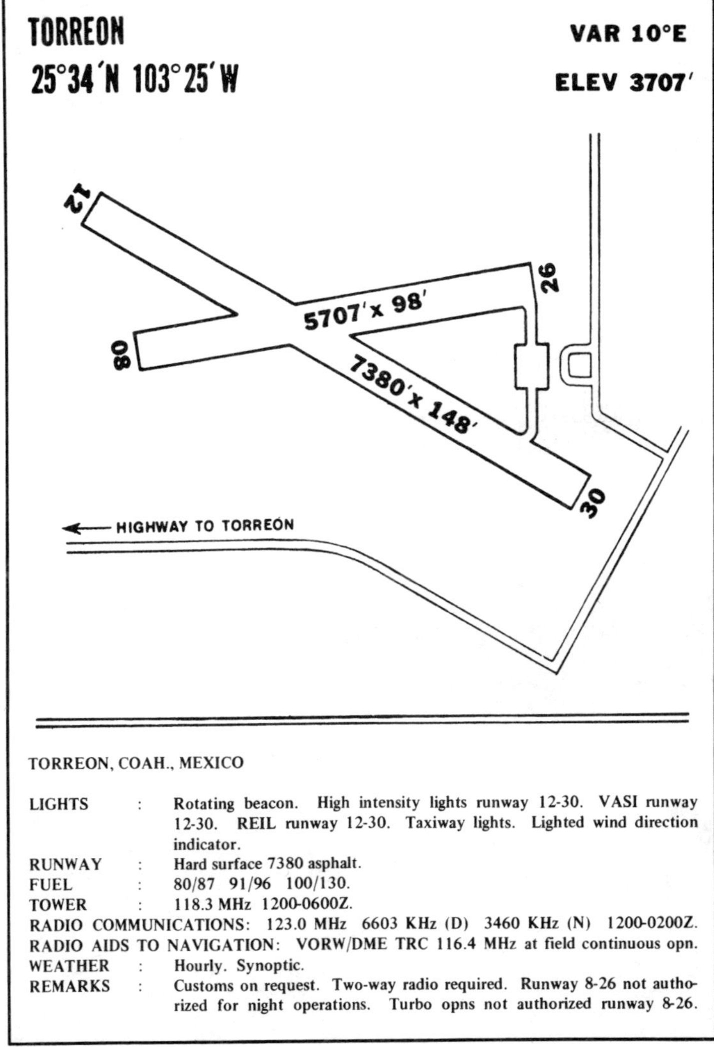

TORREON, COAH., MEXICO

LIGHTS : Rotating beacon. High intensity lights runway 12-30. VASI runway 12-30. REIL runway 12-30. Taxiway lights. Lighted wind direction indicator.
RUNWAY : Hard surface 7380 asphalt.
FUEL : 80/87 91/96 100/130.
TOWER : 118.3 MHz 1200-0600Z.
RADIO COMMUNICATIONS: 123.0 MHz 6603 KHz (D) 3460 KHz (N) 1200-0200Z.
RADIO AIDS TO NAVIGATION: VORW/DME TRC 116.4 MHz at field continuous opn.
WEATHER : Hourly. Synoptic.
REMARKS : Customs on request. Two-way radio required. Runway 8-26 not authorized for night operations. Turbo opns not authorized runway 8-26.

Fig. 8-4. Sketch of the Torreon airport. Fast fuel facilities were available here many years ago. They may still exist.

Fig. 8-5. Town square and bandstand in Durango. This area is off the beaten path for tourists, but it is an interesting place to visit (courtesy Mexican National Tourist Council).

no facility for closing the plan at that time. After topping off with fuel, we headed up the main highway to Guadalajara, the second largest city in Mexico (Fig. 8-7). The sprawling Don Miguel Hidalgo International Airport is miles east of town, so we picked up a rental car to take a tour of the

Fig. 8-6. Mexican "guard" sleeps under the new Cessna on the grass flight strip (since paved) at San Blas.

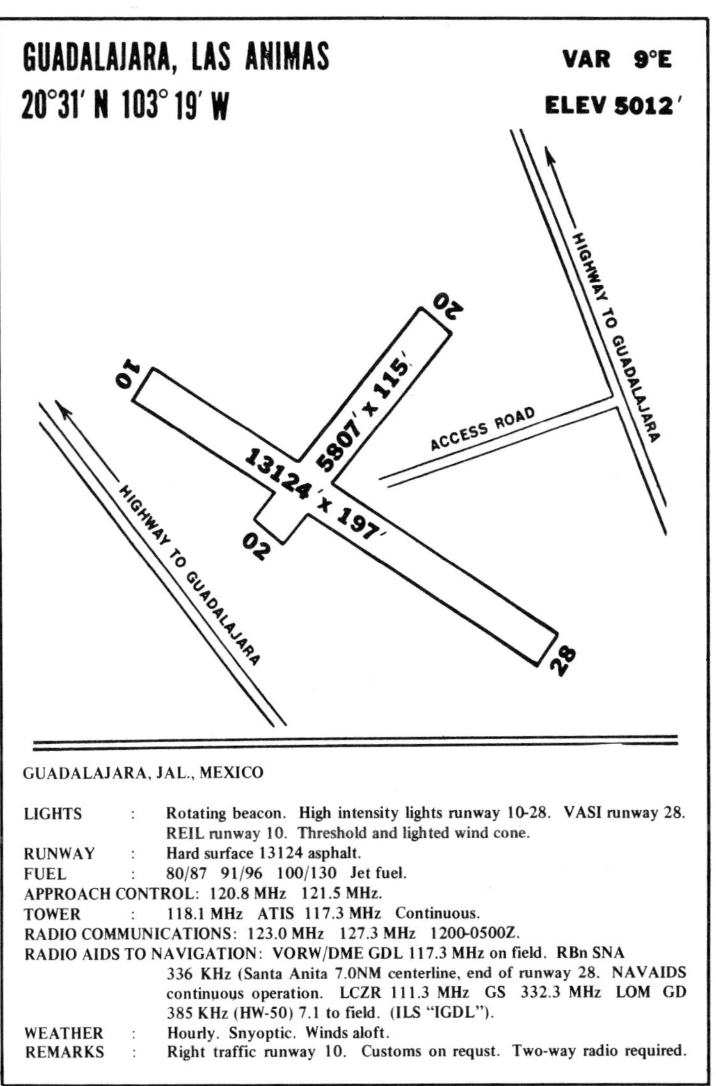

GUADALAJARA, JAL., MEXICO

LIGHTS : Rotating beacon. High intensity lights runway 10-28. VASI runway 28. REIL runway 10. Threshold and lighted wind cone.
RUNWAY : Hard surface 13124 asphalt.
FUEL : 80/87 91/96 100/130 Jet fuel.
APPROACH CONTROL: 120.8 MHz 121.5 MHz.
TOWER : 118.1 MHz ATIS 117.3 MHz Continuous.
RADIO COMMUNICATIONS: 123.0 MHz 127.3 MHz 1200-0500Z.
RADIO AIDS TO NAVIGATION: VORW/DME GDL 117.3 MHz on field. RBn SNA 336 KHz (Santa Anita 7.0NM centerline, end of runway 28. NAVAIDS continuous operation. LCZR 111.3 MHz GS 332.3 MHz LOM GD 385 KHz (HW-50) 7.1 to field. (ILS "IGDL").
WEATHER : Hourly. Snyoptic. Winds aloft.
REMARKS : Right traffic runway 10. Customs on requst. Two-way radio required.

Fig. 8-7. Sketch of the large jetport at Guadalajara. This is the second largest city in Mexico with an altitude of 5012 feet.

delightful city. Here we used a U.S. credit card for identification but not for billing. We were billed anyhow and despite copies of the signed receipt, it took a year to get that charge eliminated by the credit card company.

GUADALAJARA

Guadalajara has something for just about everyone. When it was conquered by Cortez' soldiers, Guadalajara was named after a town in Spain which, in Arabic, means "rock-strewn riverbed." According to the National Geographic Magazine, it has the second best climate in the world with a year-round temperature of 75°.

Things to see and do are as varied as your interests. *Guide* magazine, whose main offices are in the city, explains it this way:

"Guadalajara is a city of roses and romantic charm, of flowers like the colorful bougainvillea and the delicate jacaranda, fountains and colonial monuments. In the modern section, Zona Chapultepec, which is Guadalajara's Zona Rosa, there are broad tree-lined avenues surrounded by good restaurants, elegant shops, and galleries. While in this part of town, you should visit the museum which was once Orozco's house and studio, and still houses a number of his works. For shopping, you can stay in this area, or go to nearby Tlaquepaque or Tonala for traditional glass and pottery. Twenty distilleries await you at Tequila, which gave its name to the fabled Mexican beverage. But good spirits abound in Guadalajara whose people are noted for them. The Mexican national costume, the *charros* (or gentlemen riders), the typical musicians, the mariachis, and the Mexican Hat Dance all originated here in the land of the Tapatios."

If you are a traveler who, when in Rome, does it like the Romans, then you'll want to investigate the Tequila bottled in this, the State of Jalisco. The largest and most modern distillery is Tequila Sauza located in Vallarta 3273 in Guadalajara. They have guided tours with English speaking guides. If your airplane is securely tied down for the night, you might want to explore one of the popular Mexican concoctions whose recipes are listed here:

Margarita

1½ oz. of Tequila
½ oz. of Contreau or Triple Sec
½ oz. of lemon juice or lime juice
Mixed in crushed ice and serve in champagne glass rimmed in salt.

Tequila Sour

1½ oz. of Tequila
¾ oz. of fresh lemon juice
1 teaspoon powdered sugar
Shake with crushed ice and strain into sour glass. Garnish with lemon slice and cherry.

Tequila Sunrise

1½ oz. Tequila
4 oz. of orange juice
1 teaspoon Grenadine
Serve with ice cubes in a tall glass.

Ruso Negro (Black Russian)

1 oz. Tequila
1 oz. Kahlua
Serve over rocks.

Medias de Seda (Silk Stockings)

½ oz. Tequila
½ oz. Cream of Cacao
⅓ oz. of Grenadine
1¾ oz. of evaporated milk
2 cubes of ice
Mix in blender and serve in champagne glass with sprinkling of cinnamon and one red cherry.

While Guadalajara isn't quite as far above sea level as Mexico City (5012 vs. 7341), there is still an altitude adjustment factor for that throttle-to-bottle rule.

Considered as part of Guadalajara, but actually five miles east of the city on route 80 is Tlaquepaque, an area famous for its handpainted pottery, Indian weavers with hand looms, glass factories, and artisans working with silver and copper. If you want to purchase "Made in Mexico" products, Tlaquepaque is recommended as the place to buy them. The price can be right if you are prepared to bargain.

After takeoff from Guadalajara we skirted Lake Chapala, the largest lake in Mexico, and watched colorful sailboats on the calm surface. Then we climbed to 12,000 feet over the *barranca* (canyon) country between Morelia and Mexico City. Eventually the Mexico City tower operator replied in perfect English, "Cessna 3109 Romeo, report over the tall buildings downtown for a left base to Runway 5 left."

Our approach from the west was between two peaks, 12,385 and 12,917 feet, respectively. We were able to locate the massive 80,000-seat Estadio Azteca that was to house many of the events of the Olympic Games.

As we approached the airport, we could see a large jet turning around at the end of the 10,236-foot runway in preparation for takeoff. It appeared that he would probably be braking ground at just about the time we would be touching down, so we requested the right runway and were cleared to land.

Our notes from this 1968 flight bemoaned the small scale, 1: 1,000,000, ONC charts used for visual navigation. We saw charts on a 1:250,000 scale, similar to our sectionals, at that time but distribution was limited to U.S. and Mexican military pilots. While there was no classified information on these large scale charts, they were unavailable to visiting pilots. The same situation existed during our most recent flight into Mexico where just three 1:1,000,000 WAC charts cover virtually the entire country.

On this pre-Olympic games trip, we found that a *Pilot's Guide to the HemisFair and Mexico* accompanied by an excellent map were then available without charge from the State of Texas Aeronautics Commission. Even today, the Mexico Flight Manual is available without charge from the Texas Aeronautics Commission, P.O. Box 12607, Capitol Station, Austin, TX 78711.

During this visit we teamed up with Antonio Silva, Aviones S.A. pilot with a turbo Cessna 206 to shoot air-to-air pictures near 17,887-foot Mt. Popocatepetl ("Popo") (Figs. 8-8, 8-9). After the photo session, all radio communication from the T206 was lost so we followed him down to a landing on a hardpacked dry lake then used as a touch-and-go location. The T206 had suffered a complete electrical failure so the Mexican pilot climbed into our ship and handled the radios on the short hop back to the Capitol's airport. It was my poorest landing on the entire 41-hour trip.

The temperature was 80°F when we took off from Mexico City at 1:30 p.m., producing a density altitude of 10,400 feet. However, the then-new Skylane had good performance and we were airborne easily. We had planned to go directly to Ciudad Victoria, but made a detour to see the pyramids and then encountered some unforecast cloud cover enroute. We changed our flight plan to top the weather at Tuxpan and continue up the coastline to Tampico.

We could have made Matamoros and crossed into Texas with ample fuel, but we decided to land for a breather at Tampico. On this 1968 flight that decision proved to be a mistake. What followed was the only paperwork problem I've ever had in Mexico. At Tampico we were met by the Airport Comandante and the chief of Customs who immediately asked for our tourist cards and the paperwork on the airplane. When the yellow H.D.R.F.A. Form #28 was not in the package of papers, things began to get sticky. I finally convinced the officials that I had never been issued that particular piece of paper when entering Chihuahua. When I asked for fuel to continue up the coastline, I was informed that the truck was in town, and besides, I didn't have time remaining to make Matamoros by 6:00 p.m.

Fig. 8-8. Snowcapped peak south of Mexico City stands up through a ring of clouds. Mt. Popocatepetl ("Popo") is 17,887 feet above sea level and is perenially snowcapped.

Fig. 8-9. Research Cessna circles a new utility stadium prepared for the Mexico City Olympic Games in 1968 (courtesy Dana Downie).

Even with the rare commodity of a 30-knot tailwind, I was unable to convince anyone to allow me to leave that day. However, in all fairness, I checked later with the U.S. Consul, himself a pilot who happened to be at the airport, and he advised that the airport officials had been instructed to be particularly cautious because of recent accidents caused by darkness and/or marginal weather.

The next morning we were off for the border. We advised Brownsville FSS of our ETA and landed at Matamoros to clear out of Mexico. Everything went fine and we were walking out the door of the Matamoros terminal when the Customs man asked for "that yellow paper" which we didn't have. One thing led to another and the Official advised that we'd been flying illegally in Mexico for the past ten days and were liable for a fine, aircraft confiscation, or whatever. It was a rather uncomfortable few minutes and it took considerable talking before we were given a green light to re-cross the border.

It should be noted that this problem occurred back in 1968. Since that time, the Government of Mexico has gone to great effort to cut down on the pieces of paper and make flying tourists more and more welcome.

Chapter 9
Mexico City

Mexico City is a big, busy, sprawling cosmopolitan center of up to 15 million people, depending on how far out you define the city limits. Residents call the area "probably" the biggest city in the world. It has its own brand of smog generated by an endless procession of taxis, ill-tuned trucks, more old buses than new ones, and its location in a natural bowl. The 570-square-mile area is called the *Vale de Anahuac* (Valley of Anahuvac).

If you're flying yourself to Mexico City, chances are that you'll mix it with the big boys and go into the main International Airport (Fig. 9-1). There is approach control, radar, ATIS, tower, ground control, and all that big-town stuff. Excellent general aviation facilities are located on the southeast side of the field, with ASA fuel pumps located on the ramp in front of the compact but complete general aviation terminal. Weather, flight plans, customs, in-flight lunches, and everything you need is available here.

During your travels while flying yourself around Mexico, you may find it practical, either because of weather, mechanical problems or economics, to do a part of your traveling on Mexican airlines. The present U.S. stringent rule of no passengers in the cockpit does not apply in Mexico to date. If you can contact a member of the flight crew before takeoff, and if their time permits, there's a good chance that the Captain will permit you on the flight deck.

We rode most of the way down the west coast of Mexico in the cockpit of a brand new Mexicana 727 on our last trip to Mexico City. The route looks just a bit different from 33,000 feet than it does from 500 to 5000 feet in our Cessna 170.

Landing at Puerta Vallarta on a scheduled stop, we were a little surprised that the runway, impeccably smooth in our lightplane, produced a series of bumps for the heavy air carrier that crossed the approach end of the runway at 138 knots.

From this same vantage point, the valley of the City of Mexico and its snowcapped peaks is easy to see. The smog layer, so frequently associated

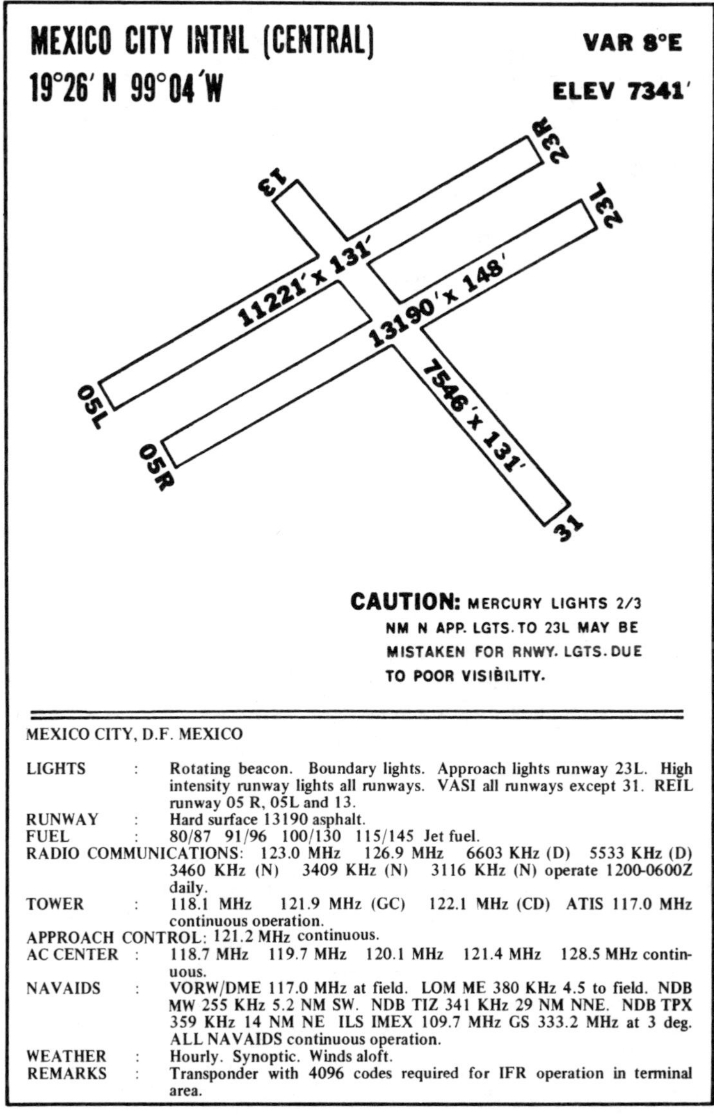

MEXICO CITY, D.F. MEXICO

LIGHTS	:	Rotating beacon. Boundary lights. Approach lights runway 23L. High intensity runway lights all runways. VASI all runways except 31. REIL runway 05 R, 05L and 13.
RUNWAY	:	Hard surface 13190 asphalt.
FUEL	:	80/87 91/96 100/130 115/145 Jet fuel.
RADIO COMMUNICATIONS	:	123.0 MHz 126.9 MHz 6603 KHz (D) 5533 KHz (D) 3460 KHz (N) 3409 KHz (N) 3116 KHz (N) operate 1200-0600Z daily.
TOWER	:	118.1 MHz 121.9 MHz (GC) 122.1 MHz (CD) ATIS 117.0 MHz continuous operation.
APPROACH CONTROL	:	121.2 MHz continuous.
AC CENTER	:	118.7 MHz 119.7 MHz 120.1 MHz 121.4 MHz 128.5 MHz continuous.
NAVAIDS	:	VORW/DME 117.0 MHz at field. LOM ME 380 KHz 4.5 to field. NDB MW 255 KHz 5.2 NM SW. NDB TIZ 341 KHz 29 NM NNE. NDB TPX 359 KHz 14 NM NE ILS IMEX 109.7 MHz GS 333.2 MHz at 3 deg. ALL NAVAIDS continuous operation.
WEATHER	:	Hourly. Synoptic. Winds aloft.
REMARKS	:	Transponder with 4096 codes required for IFR operation in terminal area.

Fig. 9-1. Sketch of the Mexican City International Airport. Elevation is 7341 feet. General aviation facilities are on the east side of the field.

Fig. 9-2. Turbo Commander taxis in at the General Aviation Terminal at Mexico City. Corporate jets and light aircraft can be seen in the background.

with Los Angeles, is visible for 50 miles on an otherwise clear day. The Mexican crew was flying in barely VFR conditions and elected to fly a full IFR procedure with an ILS approach to the runway. As we breathed down the Captain's neck, the airport slowly materialized out of the murk and we were again in Mexico City.

While the airline side of the airport handles a great many people in a remarkably efficient manner, it's my personal belief that it is much more enjoyable to taxi your own airplane up to the Nacoa pumps, top off, tie down, walk through the small general aviation terminal, and not have to worry about reconfirming an airline ticket (Fig. 9-2). One travel bureau desk in the Reforma Hotel had the following sign permanently displayed. "No matter what anyone tells you, confirm your return flight now!" Airline reservations are automatically cancelled if not reconfirmed at least 48 hours in advance. So smile all the way back to your own little airplane.

There is no inside-the-fence surface transportation to the northwest airline side of the field. Because of street and subway (the Metro) construction, it's a 30-minute cab ride from one side of the field to the other (Fig. 9-3).

A taxicab stand outside the general aviation terminal is your only way of getting to town. Cab fares are very reasonable and one of the few bargains in Mexico City. Closest of the modern hotels to this area is the Aeropuerto Holiday Inn, Blvd. Puerto Aereo 502, with all stateside amenities and rates equal to all but a few of the most posh downtown hostelries.

You'll find a chain of "Wings" restaurants on the International Airports at Guadalajara, Monterrey, Tampico and Acapulco. One of the more interesting Wings restaurants is located just across the street from the airline terminal at the Mexico City International Airport, No. 486 Blvd. Puerto Aero. A portion of this eatery is a cafeteria located inside an old four-engine Canadian North Star airliner. The airliner is parked in a lot adjoining the restaurant and next to the Airport Holiday Inn.

There are some 45 deluxe and first-class hotels located along the majestic Paseo de la Reforma, Avenida Juarez, and in the fashionable Niza section, so pick a phonebook, a guide book or the latest issue of the English-language *Guide* magazine and take it from there. If you phone from the airport for reservations, don't use Spanish unless you are fluent. Say "Hello" in English and the larger hotel operators will switch you to a multi-lingual desk clerk.

The International Airport is presently the only landing facility within easy distance of downtown. The general aviation airport at Atizapan, with services by ASA, is listed on the latest charts as 4200 feet long at an altitude of 8000 feet, but from the air it looks longer than that. Bus service is available to downtown Mexico City with the trip taking from half an hour to an hour, depending on the traffic.

It is always prudent to check carefully at your last point of departure before planning to land at any but the largest airport in the area. If you choose Atizapan, we'd recommend listing your alternate as the International Airport.

The Santa Lucia military airport near Zumpango north of Mexico City has been discussed as a future general aviation terminal, but nothing has been finalized as this book is prepared to open this facility to civilian aircraft. Note the MMR 100 restricted area, by NOTAM, to 10,000 feet at Santa Lucia but the surface elevation is over 7000 feet in this general area.

An interesting rule of thumb for flying in and out of Mexico City was

Fig. 9-3. Exterior of the General Aviation Terminal in Mexico City. All flight planning, weather, and packaged goods are available here. Cabs are available from here to downtown.

Fig. 9-4. Pino Suarez station on the Mexico City "metro" (subway). This people mover is efficient, clean, quiet and very reasonable.

provided by Mariscal Flores who flies jets for the Mexican Attorney General. "The midday *siesta* in this part of the country is strictly a function of density altitude. Normally the density altitude of Mexico City is 10,200 feet at 4 p.m. and only 4200 feet at 4 a.m. That's why most of the heart attacks take place in the middle of the afternoon around here. Never take off for any place at 4 p.m."

THE METRO

Anyone who flies small airplanes has an interest in varied forms of transportation. After you've sampled the bumper-to-bumper surface traffic jams on the way in from the airport, take the time and effort to ride Mexico City's Metro (subway). The government is justifiably proud of this superb people-mover that is expanding year-by-year from the center of town. Soon (shortly after 1981) it will have a station out as far as the International Airport, making the trip downtown much more rapid (Fig. 9-4).

First-timers to the Metro are cautioned to keep their eyes open because you can't hear the trains coming. They run on rubber tires and the units with ten or more cars in a train will come rocketing into the stations at high speed and decelerate rapidly. Large signs announce the name of each subway station and the traveler who plans in advance might consider looking at the layout of the subway available at each station and count the number of stops between where you are and where you want to be. This saves last minute shoving toward the exit. The cars are clean, though crowded in the downtown areas; the stations are well-lit and patrolled by uniformed police. While some of the executives we talked with who lived in Mexico City admitted sheepishly that they'd never ridden the

Metro, we'd recommend it as an interesting experience. There are three different routes with some outlying portions running above ground.

Riding the Metro is by far the best transportation buy in all of Mexico (Fig. 9-5). You can get five tickets for about a quarter, though the actual cost per passenger is about four times the ticket cost. The Metro is government-subsidized and about two million passengers are carried daily. Originally designed to open with the 1968 Olympics, it didn't quite make the completion time. One of the many delays was the discovery of hundreds of tons of Tenochtitlan artifacts. A shrine to Ehecatl, god of the wind, was left in place at the Pino Suares station.

Riding the Metro with its shops and restaurants at all stations is a most worthwhile experience to get a look at a fascinating portion of this bustling city.

OTHER TRANSPORTATION

Deciding how to travel around Mexico City and its environs depends upon your mission and your length of stay. A personal viewpoint is to use cabs. Good rent-a-car facilities are available at the International Airport, but on the airline portion of the airport. One guide book says, "Car driving can be a little hectic for the foreigner. Street names often change from one block to another, signs may be in unfamiliar language, and traffic circles (*glorietas*) seem to many to be a game of Russian Roulette. Just remember the old saying 'It's every man for himself,' and you'll soon get the swing of it."

Fig. 9-5. Wide downtown entrance to a metro station. Downtown Holiday Inn can be seen in the background. A metro ride is recommended to all visitors to Mexico City.

Fig. 9-6. View of the National Theater as seen from the top of the 44-story Torre Latino graphically illustrated the traffic problems in downtown Mexico City. The easier way is to take a cab or a bus.

Parking places in town are just about impossible to find, a condition that makes the rent-a-car even less desirable (Fig. 9-6). Lean over the railing of the observation deck of the 44-story Torre Latino building at the corner of San Juan de Letran and Madero (Fig. 9-7). And you can see a maze of rooftop parking lots and tiny areas where only a handful of cars find shelter. Big, well-lighted parking lots or multi-decked parking areas are a rarity.

And when you're afoot downtown, groping for the proper light signals, don't blunder out into traffic. These drivers are protecting themselves from each other, but not from pedestrians. Taking pictures from the safety zone in the middle of the Paseo de la Reforma on a hazy, drizzly night is quite a soul-searching experience.

In addition to rent-a-cars and buses, Mexico City has *Peseros*, cabs that run along specific routes. These are painted white with green stripes and the driver will have his hand out the window if he has room for another passenger. *Sitio* taxis—the name is painted on the side of the cab—belong at specific taxi stands. They can be ordered by phone if you feel that you have mastered the language, or picked up on the street. Arrange for the cost of your trip in advance. In addition, there are Turismo taxis, usually found in front of hotels. Their meters are covered and the drivers can charge just as much as you want to pay, which is considerably higher than the other cabs. However, many of these drivers will speak some English, and that may be worth the extra fee.

During our most recent visit to the Capital, the government was inaugurating a set of new buses on Paseo de la Reforma with a no-fee offer that was hard to beat. Of course, the buses were jammed and sit-down space was available only at the end of the line. However, lines of waiting passengers at the bus stops were orderly and everyone seemed to enjoy the jampacked rides. We kept running into other Americans utilizing the buses for sightseeing around the central part of the city, especially visiting Chapultepec Park where the world-famous Anthropology Museum is located. We found the museum to be extremely well put together in a series of modern buildings first opened in 1964. Even if you are not an aficionado of earlier civilizations, you shouldn't miss this presentation when you're in town (Figs. 9-8 through 9-10).

Nearby is Chapultepec Castle where Emperor Maximilian and Carlota lived. The castle was began late in the 18th century on "the hill of the

Fig. 9-7. Typical downtown traffic in Mexico City. Torre Latino building (44 stories) is in the background. There is an observation deck on top and a restaurant and bar one flight down.

Fig. 9-8. Entrance to the fine National Museum of Anthropology. Even if anthropology isn't your favorite hobby, a tour of this excellent museum is recommended.

grasshoppers." (*Chapultepec* in the Aztec language is grasshopper.) Today it is a museum that still functions occasionally as the site of ceremonial functions. It's open from 9 a.m. to 6 p.m. with a small admission.

SIGHTSEEING

Another option for sightseeing is the guided tour, usually in a group

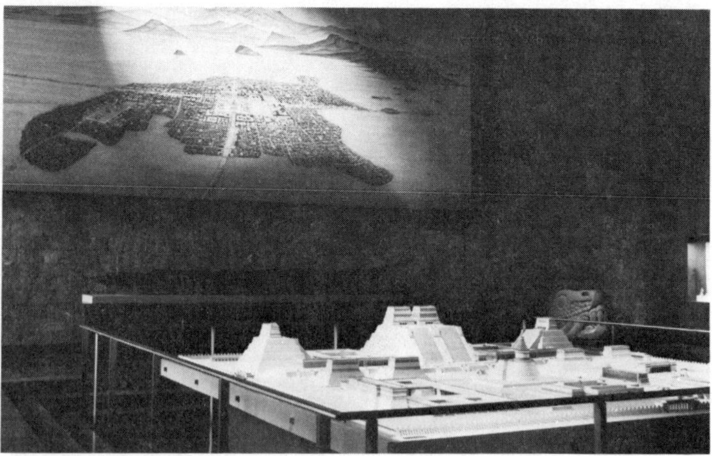

Fig. 9-9. Model of an early pyramid structure is seen in the National Museum of Anthropology. Painting on the wall duplicates the model in the foreground.

Fig. 9-10. Aztec Calendar stone on display at the Museum of Anthropology draws interested groups of students and out-of-state visitors.

and by bus. There are many available through agencies at the various hotels. We took one such tour to the Basilica de la Virgen de Guadalupe. The older church, now used as a museum, and the new circular-designed church are both located in the cobblestoned Plaza of the Americas (Fig.

Fig. 9-11. Church of the Virgin of Guadalupe is a popular visiting destination for Mexicans and foreign tourists alike. Atop the hill in the background is the older church, now a museum.

Fig. 9-12. Interior of the new church of Guadalupe presents a contrast in architecture. This time exposure shows the ornate lamps near the ceiling of this domed building.

9-11). The contrast in architecture is awesome and the new structure, without being irreverent, looks a little like the interior of the Super Bowl (Fig. 9-12). Overlooking both buildings on Tepeyac Hill is the Shrine of Guadalupe.

We also visited the Piramides de San Juan Teotihuacan on the same tour and enjoyed very much leaving the driving to someone else as well as an English-speaking tour guide and were only sorry we had not visited the Anthropology Museum before the pyramids (Fig. 9-13).

Mexico City houses the largest bullring in the world in the Plaza Mexico, seating 50,000. Bullfights are staged on many Sundays of the year at 4 p.m.; however, winter is the big season for bullfights. If you've never been to one, it is suggested that you use the guided tour for your first visit as even buying a ticket and finding a seat is no easy chore for the novice. The biggest challenge can be getting back to your hotel—there is a singular lack of empty taxis!

It's practical to have some idea ahead of time where you want to go and what you want to see (Fig. 9-14). Use an Auto Club travel guide (the AAA books are updated annually and are excellent) or one of Stephen Birnbaum's guide books. A detailed *Travelers Guide to Mexico* is found in many hotel rooms and can be purchased. The *Guide* magazine, distributed free at most hotels, has issues covering Mexico City, Acapulco, Mazatlán, Puerto Vallarta-Guadalajara-Manzanillo, and the Tijuana-Ensenada-Mexicali areas. Editor Beldon Butterfield is most enthusiastic about the Mexico City area. He advises, in part:

"Mexico City. Crowded. Car madness. The best food and shopping in Mexico. Indoor museums and outdoor pyramids. Murals. History. People parks and huge stadiums. High rises and Spanish colonial buildings. A valley surrounded by the most beautiful countryside and weather in the world. Modern day madness in a world so very different from yours. Contrasts. Savor it. Enjoy it. Live it. You won't be sorry. Scared of getting sick from the food? Don't be. Most people stay healthy in Mexico. Most who get sick do so from eating and drinking too much, too fast, too quickly. You're at seven thousand feet, so why not give your body a chance to adjust while you let your eyes feed on this uniqueness called Mexico? Scared of getting mugged? Don't be. The word does not even exist in the Spanish language. This is a big city, but it also a safe one. Afraid nobody will understand you? Forget it. Most people have some working knowledge of English and sign language does the rest. Get out of your hotel and explore."

Interesting advice. Frankly, we don't go exploring on our own at night in cities like Los Angeles, San Francisco, Chicago, New York, or Washington, D.C. in our country, much less in a foreign country. If you're with your own group or on a tour, all well and good. We believe this philosophy is applicable in any country in the world that we've visited, and we've so far covered every continent, collectively or singly, except Anarctica.

There's a magic telephone number in Mexico City, courtesy of the Mexico Tourism Secretariat. Tourists who have a problem or a question can call 250-0123 for "a helpful, friendly and courteous answer" in English.

DINE AND DANCE TIME

No visit to Mexico would be complete without sampling true Mexican cooking. Under the notation, "Try It, You'll Like it," *Guide* magazine details some of these gastronomic adventures, as follows:

Fig. 9-13. The Pyramides de San Juan Teotihuacan are most impressive and within an hour's drive of Mexico City. Tours to this area are available from any local hotel.

Fig. 9-14. Time exposure of the downtown Paseo de la Reforma Avenue as seen from the top of the Reforma Hotel. Night life in Mexico City goes on and on . . . and on.

For breakfast you might try any one of the many exotic fruit juices followed by *huevox Mexicanos* which are scrambled eggs with onion and little tomatoes known as *jitomatoes*, and chopped fresh green *chiles jalepenos*, usually kept within reason for the sensitive stomachs of most tourists. Or you can try *huevos rancheros* which are fried eggs on a tortilla topped with chili sauce; or *machaca* which is a spicy beef jerky made with scrambled eggs. With your eggs you can have little French bread-like rolls called *bolillos*, or you might try wheat or corn tortillas, and to drink, the delicious milky coffee, *cafe con leche*, or plain tea or herbal tea.

Mexican dishes are a blending of pre-Hispanic and Spanish recipes. Corn is the base for most Mexican cooking. The kernels are soaked to soften them and then ground into a *masa* which is the dough from which the tortillas are pressed and then baked, usually on a clay or metal griddle. The tortilla dates from the pre-Hispanic era and is thin and pancakelike. It becomes an *enchilada* when filled with meat and onion or cheese and cooked in a tomato and chili sauce. It is a *taco* when rolled and filled with meat, eggs, chili, vegetable or cheese. For the fried U.S. style taco, you must ask for *tacos dorados*, or for the larger sized *flautas*. *Tostados* are open-faced fried, crunchy tortillas with meat or potato and shredded lettuce or cabbage and a dash of cream on top. *Chilaquiles* are chopped tortillas stewed in a chili sauce. Tacos are usually served with fried mashed beans and cheese, onions, tomatoes and *guacamole*, which is mashed avocado with chopped onion and tomato. In northern Mexico, *burritos* are made from wheat tortillas and are filled with meat or machaca. You can order *queso fundido* which is a plateful of melted cheese, like fondue, and served with tortillas on the side. *Tamales* are made of cornmeal mixed with meat, or sweetened with sugar, and served in a sheaf of corn. When you see *lonche* on the menu, it means a sub sandwich made with bolillos. These, when toasted, are called *tortas*.

Not all chilis are hot. They vary in size, color and spiciness. An interesting sauce used in the dish *chamorro*, which is leg of pork served in a banana leaf, is *mole*, a mixture of chocolate, chili, and spices.

To accompany your meals there are many excellent Mexican beers and wines. In the beer selections, there are light lagers like Bohemia, Superior, Carta Blanca, Corona, Tres Equis, and Tecate; in the dark types, Dos Equis, Negra Modelo or Nochebuena (found only at Christmas time). There are many varieties of good red and white wines, especially from the Baja California region. You should try tequila at least once. It is made from the fermented juice of the maguey plant. You might also try pulque or mescal, also from the plant. Gulp it down and then bite into a lime with salt. For some reason or other, salt calms the burning sensation. Tequila is an acquired taste, but once acquired . . .

For after lunch or dinner, try the aromatic *cafe de olla*, served in clay jugs. Top off your meal with one of the many types of coffee liqueurs Mexico has to offer. *Bon appetit*!

Mexico City, like other cosmopolitan centers, makes a production out of dining. Their big meal is usually in the middle of the day, and they start late. Restaurants don't usually open for "lunch" until 1 p.m., and frequently there won't be anyone there before 1:30. However, downtown cafes and the ever-present fast food establishments (yes, there is a McDonald's franchise in Mexico) keep business hours that visitors are accustomed to using. We had one appointment at the airport set for 4 p.m., and the Mexican businessmen were even a little later than that returning from their "lunch."

Dinner is much the same. A Mexican pilot took us to dinner at a fine French restaurant, Rivoli, in the Zona Rosa. By the time he had solved the parking problem, we walked into the dining room just before 9 p.m.; there was hardly anyone seated. However, 30 minutes later, the room began to fill rapidly and by the time we departed to make a flight early the next morning, there was a waiting line of "beautiful people" in elegant clothes.

Continuing its truly cosmopolitan flavor, Mexico City has a night life all its own (Fig. 9-15). This night life starts late and goes later with 10 p.m. a good time to arrive at an elegant dine-and-dance spot. Reservations are recommended. The shows don't begin until around midnight, even though they may be scheduled for earlier. For your first excursion in the capitol at night, it might be well to remember again that, because of the altitude, one drink packs the punch of two.

Browse through a list of nightclubs and you'll be tempted with phrases like "An intimate meeting place . . . a spectacular view, dancing to two live orchestras . . . a jet set disco open 9 p.m. to 3 a.m. . . . psychedelic lighting—terrific for people watching, especially if you're young at heart with good ears, open 9 p.m. to 4 a.m. Reservations a must . . . this is swinging, both the crowd and the music . . . the famous violins are the attraction here . . . a totally delightful folkloric show . . . mariachis keep the tempo warm, and good company goes along with good drinks . . . a jumping night club with modern dancing, great music, and people watching from 9 p.m. till 4 a.m. . . . a bar for people who don't like bars; clublike, comfortable atmosphere with a female d.j., backgammon, good drinks and large screen TV with programs from the States . . . its guys and gals perform the latest disco beat. A place for young people with money to enjoy it . . ."

SHOPPING

If your world travels have covered both Paris and now Mexico City, you'll find a singular similarity between Mexico City's main street and the

Fig. 9-15. Lighted cathedral in the square in Mexico City at night. Spectacular lighting makes this building an outstanding sight (courtesy Mexican National Tourist Council).

Fig. 9-16. Statue and traffic circle on Paseo de la Reforma make the intersection of three two-way streets. During morning and evening hours, the traffic here can be nothing short of terrific (courtesy Mexican National Tourist Council).

one in Paris. And well there should be since the Paseo de la Reforma was ordered built by Emperor Maximilian as a copy of the Champs Elysees (Fig. 9-16). It is considered the most spectacular avenue in the city and is growing every year as more and more towering commercial and hotel buildings are completed. Sidewalk cafes make a fine rest stop for the feet and provide a vantage point for people watching.

While we didn't visit it with the idea of shopping, one of the more interesting down-to-earth spots in Mexico City is the Mercado la Merced, the largest food and curio market in the city (Fig. 9-17). There are more than 3000 vendors sprawled over several blocks with both wholesale and retail sales, but it is basically a wholesale market. Handicrafts are to be found in the east side of the market. The Merced Market is easily reached on Metro Line #1 with an exit right at the Merced Station.

One of the singular disadvantages of a lightplane trip is the lack of space and weight allowance for extensive shopping. That's not to say that you can't drop a big bundle on jewelry or perfume that comes in very small packages. Shopping here is an art in itself. We watched visitors on tour buses purchase curios from vendors at the pyramids, discuss their "bargains" on the return trip and then find similar items at half the price in downtown stores. If you admit that you don't know the value of local handicrafts, stick to purchasing in established stores like Sanborns (21 Sanborns stores in Mexico City alone) or the government sponsored FONART, with 19 stores throughout the country, featuring fixed, reasonable prices. Be sure to ask for receipts so you can prove the value of your

Fig. 9-17. A section of one of the sidewalks adjoining the Mercado la Merced, the largest food and curio market in the area. Here 25 personalized Christmas cards will cost you 40 Pesos.

purchases when returning to the U.S. Some of the best buys here are in native gold and silver artifacts, copper, onyx, leather, and textiles. You can buy pottery, woodwork, or blown glass, but these items won't fit in the normal small airplane in any quantity.

Fig. 9-18. The National Cathedral in Mexico City is an outstanding example of architecture that is a series of buildings of which the Mexicans are justly proud. Note scaffolds erected to repair the left tower of this church.

A visit to Mexico City (Fig. 9-18), regardless of the length of time involved, can be nothing more than a sample, just like one jelly bean in a candy store. If your personal desires run toward the big and busy metropolis, then you'll love this city. If you prefer the smaller, quieter places, a visit to Mexico's capital is still an eye-widening experience. When you take a taxi back to the general aviation terminal, take care of the increasingly simple paperwork and study your density altitude computations before boarding, you'll have the unique experience behind you of a visit to perhaps the largest city in the world. After takeoff from Mexico City and climbing out of its broad valley, it's literally downhill to almost any other part of the country.

Chapter 10

West Coast to Puerto Vallarta

Flying the west coast of Mexico is perhaps the most colorful of all routes in the country. It provides a true tropical destination within the range of almost every general aviation airplane. While east coast pilots may tout the fine Bahamas flight because of its proximity, there is a fair amount of flying involved. The west coast of Mexico has miles and miles of broad beaches, good roads along most of the route, and vastly improved nav/aids.

From a purely personal preference, our most popular long flight into Mexico is down the west coast. It also covers the largest coastline by far as well as well-populated areas with a myriad of tourist facilities. Thus we have divided this material into two chapters, with Puerto Vallarta as the dividing point.

On shorter trips from our southern California base, Baja, California (Chapter 12) tops the list. If you can take a week, ten days or two weeks, then we'd recommend the west coast trip for an eye-opening vista of coconut palm plantations, exotic new condominiums and high rise hotels, isolated tropical hideaways and some fine "one star" hotels with lazy Casablanca ceiling fans.

During research for this book, we took the family Cessna from Los Angeles to Acapulco and returned via the northern half of Baja. A report on this survey flight was printed in *Aero* Magazine by Editor Dennis Shattuck and is reprinted with his permission. We have added considerable detailed information on the various stops that was not used in the original *Aero* publication because of space restrictions.

Our trip took eleven days with 28½ hours flying in Mexico and another 3 hours to reach the airports of entry and departure at Mexicali and Tijuana.

Tourist cards were obtained in advance from local auto club offices.

These cards also can be obtained at any of the international airports in Mexico. A U.S. passport (valid or outdated) or a certified copy of your birth certificate is the best identification to carry with you. The complex aircraft manifest required in years past has been eliminated. Our single-sheet flight plan form, filled in with the names of passengers and signed by both the Airport Comandante and *Aduana* (Customs) at Mexicali, was adequate for the entire trip. You retain this first flight plan and turn it in when departing from Mexico.

Long gone was the RAMSA card for communications services a special fee for weather information and landings. All these are now included with your fuel bill. However, aircraft liability insurance is required in Mexico. This cannot be purchased at most international airports, requiring a trip to town. It is suggested that it be arranged for in advance. We used a MacAfee & Edwards Insurance credit card which we activated by phone before leaving home. It was accepted without question. The coverage fee for eleven days on our single-engine aircraft was $19.10. You will also be required to show your aircraft registration. If the aircraft is borrowed or rented, a notarized affidavit giving you permission to fly the aircraft into Mexico is required.

Among other insurers of aircraft flying to Mexico is Mexair, which can also provide hunting and fishing licenses, arrange for gun permits, and assist pilots with obtaining tourist cards and health certificates (if required). Call Raoul Martinez at Mexair (213) 398-5797 for assistance.

There were four of us on this eleven-day exploration. Paul and June Crawford, relatives from Santa Barbara, California, are 70 years old—going on 40. Paul has shared some of his thoughts as a passenger on trips into Mexico in Chapter 3. The Crawfords have joined us on many junkets over the past years. They enjoy new places with new foods and new labels on beer bottles once the day's flight is over. They never seem to worry about the flying nor get upset with the paperwork and delays. In short, they're ideal companions on a trip, foreign or domestic.

We flew N2672D, a durable old 1952 Cessna 170B. Weather was no problem on this trip except coming back into Brown Field and San Diego's new unpopular TCA. It was very hot and humid south of Puerto Vallarta, but this trip was made during the off season. From May 15 until mid-December, most hotel prices are down as much as 40%. Runways that are 9000 to 11,000 feet long and field coastal elevations of perhaps 10 feet ASL make a 100° density altitude no great problem.

We picked our stops for a leisurely flight with 3½ hours in the air the longest in any one day. If you cross the border and relax, take what you consider delays in stride and don't build up your blood pressure, you'll have a much more enjoyable time in Mexico. Remember that you're a guest in a foreign country and try to fit in with a somewhat different pace of living.

The government-controlled ASA (Aeropuertos y Servicios Auxiliares) operates 48 *aeropuertos* in Mexico, 28 of which are international (Fig. 10-1). Ing. Javier Garcia Olave, ASA director in Mexico City, advised us that the

Fig. 10-1. ASA operations office and control tower at a U.S. Border airport. All ASA operations are available at the base of this tower in Tijuana, B.C. School children in this photograph were awaiting a free general aviation ride during "Mexican Aviation Week." More than 5000 students were given rides in a single week.

only difference between the two classes of ASA airports is that the non-international fields do not have immigration, health, and customs officials. All fuel at the ASA airports is handled by the government agency and prices are identical. Our 100-octane was 98 cents per gallon with the peso figured at 22:1. At the two private (non-ASA) airports where we fueled—Los Mochis and Santa Ines—the fuel cost was appreciably higher and credit cards were not accepted. At this writing, ASA accepts Interbank, Diners, Visa, Carnet, Bancomer, Bancam, Continental and Shell Aviation cards. However, there may be a minor hangup once in a while, as we promptly found out in Mexicali. The credit cards were fine, but the fuel office was temporarily out of charge-slip blanks. We paid cash.

In the seven other ASA fields we visited, two of them twice, we paid for all fuel by credit card. Visa was preferred over Diners. There is some delay in filling out the credit card slips at some of the more isolated fields because there is a detailed breakdown of charges that seems to give the service personnel problems.

The United States is fortunate to have both borders bridging areas of great pictorial tourist interest. The coastline down the west side of Mexico has some of the bluest water, whitest beaches, and lushest sprawling coconut palm plantations that you'll find anywhere in the world (Fig. 10-2). Isolated bays with brightly-painted fishing boats tend to coax the tourist to fly at a sightseeing altitude rather than the altitude for prime range and efficiency. We spent one evening with a couple from the San Jose, California, area visiting Mexico for the first time. Their flights had been at 8500 feet in their Bonanza, but we convinced them that a closer view of the scenery was worth the extra few minutes flight time and perhaps an extra gallon of fuel per hour. However, for each pilot there is a different comfortable cruising altitude.

With four people in a Cessna 170 you pack very lightly, but the weather is warm. We took nylon windbreakers and wore them only twice, on the first and last nights out. You can get by with a single pair of shoes and a two-day stop in Puerto Vallarta solved the laundry problem.

MEXICALI

There was a considerable delay at our Mexican airport of entry, Mexicali, but you must remember that you are now in Mexico and things may take just a little longer. It was Friday afternoon on a long weekend and there were at least 20 airplanes waiting to clear Customs, refuel, and get tourist cards checked and flight plans made. The daily Mexicana 727 also was on the ground requiring service on a turnaround, so it was a particularly busy time for the Mexicali staff.

Fig. 10-2. Broad beaches, tropical temperatures, and palm trees just south of Mazatlán on the west coast near the town of Caimareno prove attractive to Mexican vacationers.

When you plan a flight into Mexico, you really can't assume that you'll get in, refuel, clear Customs, and get back in the air as fast as you might wish. So don't plan to leave home after work, zip across the border, and have another three hours of daylight flying ahead of you. Single-engine flying at night is not approved.

Mexicali is the capital of the State of Baja California. The southern half of Baja is technically a territory. Mexicali is a large border town with population at 400,000 but since it is not on the main highway to either Baja California or into the mainland of Mexico, it is all but ignored by travel folders. Aside from a lively border trade and twice-a-month bullfights in the winter, the town is perhaps best known for its local brewery. With the trademark "Mexicali," their beer is both popular and excellent.

Since the airport at Mexicali is at least 15 miles out of town, most pilots planning to cross into Mexico here for an early takeoff will RON (remain over night) at the convenient Imperial Airport a dozen miles north of the border. The Airporter Inn on the airport here has good border accommodations with a restaurant and bar on the premises. A quarter of a mile distant, but still on the airport, is a good Mexican restaurant. The FAA staffs a 24-hour FSS and an eight to twelve-hour tower. The Imperial FSS will accept "round robin" flight plans for visitors to Mexico who plan to return through Mexicali and the adjoining U.S. Airport of Entry, Calexico. Imperial FSS has a record of maintaining one of the most up-to-date information packages on what's happening at both Baja and mainland airports. It's well worth the time of an additional landing for a nose-to-nose briefing, particularly if this is to be your first trip into Mexico.

The general aviation terminal at Mexicali is new and clean and has a nice small restaurant. The paperwork went quickly once we got through the waiting lines. To keep the first day's flying short we'd planned an overnight stop at Puerto Peñasco, a small fishing town with a VOR and a hard-surface runway located a third of the way between Mexicali and Guaymas. While the town is on the charts, nobody knew much about facilities and there was a good reason why.

PUERTO PEÑASCO

We climbed to 1500 feet in a desert haze and followed the railroad tracks over desert to the gulf. Soon we began to pick up the Puerto Peñasco VOR and we circled in over the town (Fig. 10-3). Most streets were dirt, the harbor was busy, and the airport had one airplane and one helicopter parked on the ramp. There was no tower and the airport office was closed. After surveying town, we selected the smooth dirt strip into the wind and landed. There was no pay phone and no way to call a cab. We found out later that you buzz town and a cab will come to the airport.

Eventually a shiny new pickup truck rolled into the car parking area. A mother had brought her small youngsters out to look at the airplanes. We asked for a ride into town in more-or-less Spanish, and she obliged. She finally accepted a few pesos for fuel, but did it reluctantly. There was some

Fig. 10-3. Village and harbor at Puerto Peñasco. Airport is beyond the town by perhaps two miles along the main road leading to the north. The town is home port for many Mexican shrimp boats.

soul-searching about leaving our one and only airplane tied down at a deserted airport, and we can't really recommend it either in Mexico or the U.S., but we had little option except to continue on to Guaymas that same day where there is an ASA airport. All government airports have 24-hour military or police security so pilferage is at a minimum (Fig. 10-5).

Puerto Peñasco is not high on the tourist list and really isn't the type of resort town you'd deliberately head for unless you had friends in town or staying at the Playa de Oro Trailer Resort, two kilometers east on a dirt road, where the AAA booklet lists beach, boat ramp, fishing, and laundry.

Since this recent survey trip, we were advised that an airport manager/attendant had been appointed. Whether or not fuel or other services are available is an open question that should be answered in Mexicali before departure. However, on general principles, in the smaller fields of Mexico (and Canada, for that matter), try to carry enough fuel to make it comfortably to the next big town without refueling. Then, if you *can* get av/gas along the way, so much the better.

We had no reservations, but the older Hotel Villa Granada had comfortable accommodations. During a walking tour we saw extensive shrimp boat construction, the town's number one industry (Fig. 10-4). The following morning we discovered there was no power in the town due to a sharing arrangement with Hermosillo and there was quite a wait for water and electricity for cooking. When we finally took a cab to the airport, we found the 170B safe and sound but wearing a light coating of dust from overnight winds.

Again there was no one at the airport, so we loaded up and took off. Thus we were flying with a Mexican flight plan from Mexicali to Puerto Peñasco only, but the border officials had assured us that it was okay.

Fig. 10-4. There are shipyards at Puerto Peñasco building shrimp boats for the Gulf of California. Both wood and metal vessels are fabricated in the harbor here on an assembly-line basis.

ON TO GUAYMAS

We cruised down the coastline marking off checkpoints on the WAC chart CH-22. Actually, you use only three WAC charts for the entire trip—CH22, a corner of CH-23, and most of CJ-24. You soon find a disadvantage in these smaller-scale charts, and it isn't the scale it's the infre-

Fig. 10-5. Fine airport, but "Xd" out, just north of Guaymas, was used to make the film *Catch 22*. Some of the Hollywood prop buildings are still standing. Isolated beach with white sand is popular with tourists who are driving.

quency of issuance. Our CJ-24 was dated October 1974 and was the latest chart available for the south coast of Mexico.

We used Oliver's ADF Guide that covers the U.S., Canada, and Mexico with at least one commercial station listed for each of the larger towns with airports. Our King KX-86 ADF worked well for a change, and we listened to raucous Mexican music from Hermosillo from time to time through the four Sigtronics headsets we've installed. Using a voice-activated intercom made this trip much more enjoyable for our passengers, since talking between all four people is difficult in the rather noisy cockpit environment of the 170B. With this system, everyone can hear both sides of all tower communications or listen to the ballgame on the ADF. (See the listing of Mexican broadcast stations in the Appendices.)

Kino Bay's hard-surface airport had a couple of planes in sight as we skirted Tiburon Island, known in the history books as the home of the Sari Indians who were headhunters in years past. Red desert hills climb 2900 feet and isolated bays shelter a handful of fishing boats as you near Guaymas. The hard-surface strip where the film *Catch 22* was made is crossed out, but some of the prop buildings still stand (Fig. 10-5). San Carlos Bay, with its sheltered yacht harbor, is just like the travel brochures show, and the tower operator at nearby Guaymas speaks excellent English (Fig. 10-6).

The Guaymas Airport was loaded with general aviation aircraft on this long weekend (Fig. 10-7). Tiedown ropes and "dead men" were not available, so we drove extrusions into the hard ground and used our own ropes. There was little or no wind at the time, but we remembered the night

Fig. 10-6. Sheltered yacht harbor at San Carlos Bay within five miles of the Guaymas Airport. This area is popular with U.S. tourists and retirees. The area has a large trailer park and nearby shops.

Fig. 10-7. The parking area at the Guaymas Airport was filled with visiting U.S. aircraft during a long weekend. As close as it is to the U.S. Border, Guaymas is a very popular fly-in destination.

before. Fueling was quick and credit cards were accepted—Visa is preferred over Diners.

The airport officials at Guaymas had no problem with our two-day-old flight plan from Mexicali to Puerto Peñasco. Later in the trip we had two similar hops from non-ASA fields where day-old flight plans were closed as a matter of course. So don't really plan on immediate search and rescue activities if your routine flight plan isn't closed.

Fig. 10-8. Hotel Playa de Cortez on Bocachibampo Bay is popular with visiting pilots. This hotel is relatively close to the airport, isolated from downtown noise, and has a fine restaurant and quiet bar. Rooms are large, airy and reasonable when compared with resorts farther south.

The Hotel Playa de Cortez on Bocachibampo Bay is closest to the airport and a long-time RON spot for lightplane pilots. Year-round rates are 500 to 600 pesos ($22 to $27) for an excellent two-bed room. Meals, particularly seafood, are great (Fig. 10-8).

Guaymas, like most of our subsequent stops, is an interesting town for the tourist. There are fine curio shops and a large roofed market with intriguing odors and just about anything you'd want to buy. You can charter a boat and fish, or walk around the clean streets downtown. Water skiing, horseback riding, and skindiving are popular. The shrimp docks at the south end of downtown make an interesting, if odoriferous, stop.

Guaymas is one of Mexico's finest seaports and has a colorful history. Spanish explorers first visited here in 1535 but the area was not developed until Fathers Salvatierra and Kino established a mission base here. The port was attacked by U.S. Naval forces in 1847 and occupied until 1848.

Fig. 10-9. The 18th century church of San Fernando on the public square in downtown Guaymas. The area was settled in about 1760 by Indians called Guaymenas.

Fig. 10-10. Excellent harbor at Guaymas shows up in this aerial photo. Much of the filled land along the bay is made of oyster shells that have been dumped there over many years. Large fleet of shrimp boats and odorous shrimp processing plant are located on the small promontory protruding into the bay at the right of this photo.

French buccaneer Count Gaston Raouset de Bourbon and 400 pirates attempted unsuccessfully to seize the city in 1854. However, the French under Maximilian captured the port in 1865. Supplies headed for troops in Arizona were shipped into Guaymas during the U.S. Civil War and then taken inland (Figs.10-9, 10-10).

Fig. 10-11. Large trailer park at San Carlos is only partially filled. Yacht harbor is in the background and stark redrock hills come down to meet the sea.

Fishing is excellent in this area with oyster and shrimp heading the list. Who hasn't eaten Guaymas shrimp? Sport fishing is popular with a variety of charter boats available at relatively reasonable prices. If sport fishing is your bag, Guaymas is a good place to start.

Thirteen miles northwest from Guaymas is the town of San Carlos (Fig. 10-11). It once had a fine hard dirt airport with good fueling facilities, but now it's a golf course, so you land at Guaymas and use a cab or rent car. A variety of sport fishing boats are available and there's the San Carlos Diving Center with 34-foot boat and diving guide available.

Another four miles up the coast from Guaymas is Los Algodones Beach where the films *Catch 22* and *Lucky Lady* were filmed. The airport is much in evidence; so are the broad "X" marks on each end. For some obscure contractual reason, this delightful hideaway, complete with the buildings used as "props" in these films, is off-limits to pilots. However, there's nothing that says you can't circle at a reasonable altitude and look over the area (Fig. 10-12).

Fig. 10-12. *Catch 22* flight strip on Algodones Beach looks like an ideal landing spot, but it is "X'd" out and off-limits to visiting pilots. Note suddenly-rising mountains in background and you'll see why Frank Tullman and the *Catch-22* crew made their 16-plane B-25 takeoffs out to sea.

Fig. 10-13. The 6500-foot "downtown" airport at Navajoa is the center of a large agricultural area. The primary crop is cotton. Navajoa is on the main highway between the U.S. Border and Mazatlán.

Out of Guaymas at a leisurely midday time, we followed the highway past Ciudad Obregón's broad airport and admired the paved landing strip in downtown Navojoa.

Ciudad Obregón has a VOR, a part-time tower, and two 7500-foot runways. The area is the center of the Yaqui Valley with a great deal of agriculture: wheat, cotton, rice, corn, and alfalfa. There is no colonial atmosphere and most of the large structures turn out to be grain elevators. Because of the rice fields, some of the best duck hunting in Mexico is reported here between October and February.

Nearby Navajoa has a 6500-foot paved strip right next to downtown. A new irrigation project, the Mocuzari Dam on the Rio Mayo, is some 20 miles northeast (Fig. 10-13).

If you're looking for early Spanish architecture, swing east of the main highway and railroad to Los Alamos where the newest building is over 100 years old. A federal decree prohibits any new construction or alteration of the town's colonial atmosphere. Large gold and silver deposits were discovered in the 1780s and the town grew to 30,000 people. After the 1910 revolution, Los Alamos was almost a ghost town, but American retirees began flocking to the area shortly after WW II. These new residents have restored many of the town's homes.

Alamos is also known as the home of the Mexican jumping bean. The beans contain a strong strain of larva and are picked from wild plants in the area and sold to American tourists.

While not paved, the hard-dirt Alamos airport is 4000 feet long. We visited here briefly a number of years ago in a Baja Airlines Martin 202.

LOS MOCHIS

On our survey trip, we bypassed Los Mochis on the way south and caught it on the way back; in the interests of continuity, we'll insert our experiences here. Los Mochis is the center of a growing, lucrative agricultural area growing cotton, wheat, rice, corn, tomatoes, flowers and sugar cane. A large sugar mill located upwind of the town spreads a pungent aroma over the area.

Our only poor guess for an airport for an overnight stop was Los Mochis. You have the choice of at least three airports, but high surface winds reduced our options to a privately operated airport adjoining the Holiday Inn (Fig. 10-14). Fuel was available if you waited long enough and there was a charge for overnight parking. In all our travels we had yet to get the windshield cleaned; pack your own cleanser when traveling south of the border. No phones were available at the terminal, so we climbed the hill to the Holiday Inn to get a cab: 150 pesos to town.

If you have the inclination and the time, the train trip on the Chihuahua-Pacific Railway is very worthwhile. You can purchase tickets at Los Mochis for the 18-hour, mostly daylight trip to Chihuahua City (Fig. 10-15).

When we fueled the next morning at Guaymas, the local owner of a new Maule mentioned, "I never fuel at Los Mochis." However, a new government jetport is nearing completion some 15 miles west of town so normal ASA services should soon be available.

A popular landing spot in the Los Mochis area is El Fuerte, some 50

Fig. 10-14. On final approach into a private airport at Los Mochis adjoining the Holiday Inn. At the time of this visit, the new ASA airport located closer to Topolobampo than Los Mochis had not yet been completed.

Fig. 10-15. Busy street corner in Los Mochis on a weekend. While not a tourist town, Los Mochis is the terminus for the interesting Chihuahua Pacific Railroad.

miles northeast along the railroad line to Chihuahua. El Fuerte has a hard-surface 4300-foot strip that looks very good from the air. There is a 16-unit Hotel Posada del Hidalgo near the main plaza. Excellent bass fishing and duck and dove hunting are reported.

The Lake Hildago Lakefront Lodge, seven miles north of the town of El Fuerte and 300 yards up the mountain from the east end of the dam at Lake Hidalgo, is set up for pilots. The ledge reports its own 3800-foot dirt airstrip, elevation 300 feet. The lodge reportedly checks the surface of the strip daily. When arriving, buzz the lodge and a pickup service will be waiting for your group in the parking area. Packaged accommodations, including meals, boat with guide, "bird hustler," 12-gauge shotguns, and night watchman for plane parking are available. For additional information, write Tom Jenkins or Javier (Taco) Torres, Apartade (Box) 11, El Fuerte, Sinaloa, Mexico.

CULIACÁN

We landed for fuel and a siesta at Culiacán (Fig. 10-16). The airport here is jet-sized, 7500 feet long with a VOR and excellent terminal facilities (Fig. 10-17). We watched two AeroMex DC-9s exchange passengers while enjoying a *Vita Naranja* (Orange Crush) in the restaurant. We then went back on the ramp through the "executive terminal" and filled out our fuel chit (117 liters through our 180-hp engine from Guaymas at a cost of 666.32 Pesos in 1980).

Culiacán is the capital of the state of Sinaloa and is both a mining and an agricultural center. The area grows cotton, peanuts, tomatoes and poppies that are converted to mostly legal opium. At the airport here we saw a fleet of six Cessnas that were spreading sterile flies from the air in an effort to

Fig. 10-16. Culiacán control tower. This modern airport has complete facilities for the surrounding agricultural area. Spanish language banner calls for everyone to join in a local census count.

eliminate hoards of horse flies that were killing cattle in the area (Fig. 10-18).

While Culiacán is not a seaport resort, it does have three hotels recommended by AAA.

MAZATLÁN

It's only another hour at 135 mph to Mazatlán. Along this route you pick up CJ-24, the single WAC chart that will take you the rest of the way to

Fig. 10-17. Twin-engine ambulance plane pulls into the ramp at Culiacán as an Aero Mexico DC-9 unloads baggage at the left of the photo.

Fig. 10-18. Mexican pilot shows his map for spreading sterile flies in the Culiacán area in an effort to eliminate horse flies. Parallel lines on the chart are controlled flights to dispense the flies.

Acapulco and beyond. The airport at Mazatlán is broad—7200 feet of concrete without a taxiway (Fig. 10-19). It's also a far piece from town—27 kilometers and a fixed taxi price of 250 pesos which you pay at the airport. This system is used at most out-of-town major Mexican airports to establish some control over fees charged first-time tourists. Usually the cab fare back from town is 50% of the controlled price.

Fig. 10-19. Aerial photo of the Mazatlán Airport and terminal area. General aviation aircraft are parked to the left of the tower.

When you're touring by lightplane, it's next to impossible to make reservations ahead. If you try to keep a schedule, you put yourself under added pressure, taking part of the fun out of the trip. We used AAA tour books and advice from local cab drivers. At Mazatlán, for example, the resort hotel El Cid was full, but the cabbie took us about three blocks away to a clean, adequate "family hotel" that filled the bill and didn't empty the pocketbook (Fig. 10-20).

Mazatlán is a key seaport. According to legend, pirates buried their treasures around the city. A small number of Spaniards settled here in the 1600s, but there was no municipal government until 1837. Today it is home port of Mexico's largest shrimp fleet and many shrimpers can be seen near shore on both sides of the city. Freighters, cruise ships, and ferry boats headed for La Paz and Cabo San Lucas on the tip of Baja can be seen in the busy harbor (Figs. 10-21, 10-22).

Mazatlán is both a business and a resort town. Sport fishing is excellent in waters that seldom get colder than 60° in winter or warmer than 75° in summer.

You can tour the city in horesdrawn *aranas* (spider wagons) and look at the lighthouse extended upward 500 feet and the second highest in the world. One travel brochure described Mazatlán as "an orgy of sunshine and seafood." You can take a three-hour harbor and beach cruise, but somehow this will probably lack allure since you've been cruising down picturesque beaches by the hour if you're doing your own flying.

A unique way to get around town is to use an open-air three-wheeled

Fig. 10-20. Cab driver in Mazatlán has an interesting display of decals on his glove compartment. Top left, translated roughly, reads: "When everyone says that you're a burro, bray!" Below that: "My best friend is a traitor and a very convincing liar." Next: "Ask me and we'll get married." At the far right, we're told: "A noisy muffler is against the law."

Fig. 10-21. Aerial photo of Mazatlán and its harbor. New international airport is located far from town and shows in the haze at the top right of this photo.

motor scooter that carries up to three passengers. Guess why it is called a *pulmonia.*

Mazatlán has at least eight fishing fleets, but it is still a good idea to make a reservation in advance during the busier winter months. Hunters can find pheasant, quail and dove, wild boar, deer, and wildcat; the ocelot and jaguar are endangered species. The Resort Hotel El Cid has an 18-hole

Fig. 10-22. Ships of all sizes and shapes are tied at the docks along the Bahia Darsena at Mazatlán. Ferry terminal is at the far end of the harbor.

golf course with a nine-hole course at Club Campestre de Mazatlán (Fig. 10-23).

Undoubtedly Mazatlán has some delightful beaches and clean resorts. However, the beach we walked on was littered with beer cans and broken bits of Styrofoam from impromptu surf boards (Fig. 10-24). This was the only beach we saw on the entire trip where the sands were not inviting.

But to show another view of the beaches, here's a column from Beldon Butterfield's *Guide* magazine on the subject of Mazatlán:

"The beaches here are magnificent and much preferred over the hote pool area. The water is just the right temperature and the waves are not that big, yet body surfing is a favorite pastime. In Mazatlán one has the tendency of staying close to the hotel. Going out for lunch is practically unheard of. Since life takes place on the beach and going shopping just takes you out of the sun, you will find dozens of vendors hawking everything from turtle oil to clothing, from shoes to jewelry. 'Why leave the beach?' is the attitude.

Fig. 10-23. Tennis courts and a golf course are part of the Hotel El Cid complex at Mazatlan. Other luxury hotels are located along the beach front.

Fig. 10-24. Horses along the beach at Mazatlán. Catamaran in the foreground can be rented by the hour.

"Then there is the sporting life. Ocean fishing for sailfish, the very thing that made Mazatlán famous in the first place. There is also surfing for the younger generation. But in Mazatlán one way or the other, it is always the sea." (Figures 10-25, 10-26).

"In Mexico, lunch hour starts around 2 p.m., but not in Mazatlán where 90% of tourism comes from West of the Rockies. At 1 p.m., the restaurant in your hotel is full. Afternoons are left for shopping in the many large and small mercardos, or it's back to the beach. Siesta time in Mazatlán is not an important event as it is in the rest of Mexico. Dining out is an early affair, not like Acapulco, where anything before 10 p.m. is just frowned on. By eight, Senor Frogs is full; at eleven you have the pick of the house. Going out at night is a casual affair in dress and manner. It is also a noisy affair where having a good time takes precedence over good food. You just don't dress up in Mazatlán—well, some people do.

"Mazatlán has more trailer parks than any other place in Mexico; in fact, until recently it had more trailer parks than all the rest of Mexico put together. More casual life style, more people looking towards those fabulous beaches. If you are thinking of playing golf or tennis, late afternoon is suggested; the midday sun and heat will kill you. If you want to go to the theater, see a movie, visit a museum or browse through some cozy art gallery, forget it—you came to the wrong place.

"Summing it all up, Mazatlán lifestyle is the least Mexican of all Mexican resorts, has the finest beaches and everything is casual. Life on the beach is it during the day. The nights are carefree and noisy and people do love to drink." So says *Guide* magazine!

Fig. 10-25. Picturesque view of the waterfront near a resort hotel. Tropical flowers have been planted. A shrimp boat may be seen in the harbor beyond the swimmers.

Pick up a copy of *Guide* at any hotel in Mazatlán for a long and imposing list of restaurants, night clubs and discos, shopping bazaars, and even a prospectus on buying local real estate.

Naturally, you'll pick your own restaurants. We found a fine seafood

Fig. 10-26. Broad new paved highways connect new luxury resorts with the downtown area. Puesta del Sol Hotel on the Playa Gaviotas has its tennis courts on a promontory into the bay.

Fig. 10-27. Newly-paved 3300-foot airport at San Blas is immediately adjoining the small town. People and animals cross the middle of the runway to reach houses located beyond the airport. This can make landing an experience.

restaurant, the Lobster Trap, near our hotel. The eatery has a tropical motif of lighted torches and water running over a Hollywood-type waterwheel, but don't plan on their published breakfast hours. Since we wanted to get to the airport in reasonable season, we asked and were assured that breakfast would be available at 7 a.m., *en punto.* At 7:45, two employees were beginning to clean up the place leisurely and breakfast would be available "maybe at 9 o'clock." We went a block away to a small, lively coffee shop for a quick breakfast and also saved a few pesos.

It is only 200 statute miles down the picturesque coastline to Puerto Vallarta. Coconut tree plantations line the shoreline by the mile, many of them new groves just being developed. Small, isolated villages have their covey of open fishing skiffs powered with single outboard motors. In the lagoons just inland of the shore, boats can be seen fishing with nets. Most of the mountains inland were obscured by smoke from many brush fires, some handset to clear brush land or fields while others were probably the result of lightning strikes in isolated terrains where they were left to burn themselves out.

SAN BLAS

Again, to keep the West Coast in perspective, we'll visit the colorful little town of San Blas, Nayarit, at this time. In fact, however, we used it as an overnight stop on our return trip.

We circled the village of San Blas, noted that the 3300-foot runway was now paved, and let down to land (Figs. 10-27, 10-28). The paved strip was a distinct change of pace from Mexico's jetports. Youngsters, dogs, cats, and chickens play on the runway. There is a permanent path right across the center of the strip to small tropical houses. The youngsters get their thrills by darting out in front of landing and taxiing aircraft playing "chicken."

Once on the ground and parked near a Beech Bonanza from San Jose, California, we had a crowd of a dozen chattering youngsters around the plane. They looked in with interest, but didn't touch anything. No tiedowns were available, so we checked the Cessna and took a cab to the Posada Casa Morales. Frankly, we debated about leaving our ship to the mercies of the local youngsters. However, a two-man, eight-hour-a-day military post was maintained at the airport with two uniformed soldiers on duty carrying automatic rifles. Perhaps the memory of their presence kept people from touching our unattended aircraft at night. These soldiers laboriously copied down our license number and Mazatlán destination the next morning, but did not transmit any flight plan that we knew about (Fig. 10-29).

A three-hour jungle river cruise is available. Bring your mosquito

Fig. 10-28. Ruins of a mission and fortress built by the Spaniards stand on a hill overlooking the town of San Blas. Airport is at the top right.

Fig. 10-29. Two Mexican soldiers, rifle over their shoulders, make out a record of the visiting aircraft before departure from San Blas.

repellent for this trip and for in town. From our motel we watched the long outboard-motored fishing boats depart in the evening, to return the next morning with a wide variety of fish (Fig. 10-30). The accommodations (particularly if you choose the board-and-room plan) are exceptionally reasonable. During the off season, many rooms are empty.

Overlooking San Blas are the stone remnants of a fortress and mission built by the Spaniards. At that time, San Blas was a major shipbuilding center for the Philippine trade. It is far enough off the regular tourist paths to retain a mystique that is rewarding. It is certainly no Acapulco or Puerto Vallarta, but the place is easygoing and relatively unspoiled by time.

Fig. 10-30. Small fishing boats bring in their catch at San Blas. Hotel Posada Casa Morales can be seen in the background.

Fig. 10-31. Fine jetport at Puerto Vallarta is in the foreground. Second runway, nearest the camera, was under construction at the time this picture was taken. The town is at the far right of this photo.

PUERTO VALLARTA

From San Blas, it's only another 60 miles to Puerto Vallarta. We climbed high enough over the smoky ridges north of Bahia de Banderas, called Vallarta Tower, and touched down on their ample jet runway. A parallel strip is now under construction, and the terminal area and general aviation section are modern, clean, and air-conditioned. When you top off with fuel, you get two nights free tiedown, both here and at the other ASA facilities (Figs. 10-31, 10-32).

Fig. 10-32. Terminal buildings at Puerto Vallarta. Large building in the foreground is used for passenger jet parking with the general aviation ramp on the blacktop behind the tower. Excellent general aviation terminal is just beyond the airport car parking lot.

Fig. 10-33. Posh hotels on the road to the airport. In the foreground is the Posada Vallarta with the Playa de Oro closer to the shore. The airport can be seen in the far background.

Puerto Vallarta is the exception when it comes to driving to town. The trip is relatively short and the tab was only 90 pesos for the four of us. The town retains cobbled streets which are great for local color, but murderous to drive or walk on. New, very plushy high-rise resorts line the highway between the airport and town, each with its own private pool with lanai and bar—very posh and quite expensive (Figs. 10-33 through 10-36).

Fig. 10-34. Cobblestone streets in Puerto Vallarta show through the archway of the downtown Hotel Oceano. Note statue of seahorse at the far right.

Fig. 10-35. Holiday Inn, shown in the foreground, with the Fiesta Americana farther up the beach. Note large swimming pool with island in the middle at the Holiday Inn.

On the beach at Vallarta, we talked with the owner of YoYo's, Alfredo Almaraz, who runs four boats and four 24-foot parasail parachutes on the beaches. The fee is $15 U.S. for five minutes (1980) or a little more at the end of a 500-foot towline hooked to a powerboat with a 440-cubic-inch Chrysler engine. Alfredo explained that his crews average 15 customers each day. They usually commence work just before noon when there is a five-knot wind to aid in liftoff, and they have to shut down if the wind goes

Fig. 10-36. Tropical trees and thatched roofs rim the ornate swimming pool at the Hotel Fiesta Americana.

Fig. 10-37. Alfredo Almaraz, right, and an assistant, help buckle the parachute harness on a young Mexican girl who is about to make her first parasail flight.

Fig. 10-38. Off she goes! Tow rope tightens and a parasail flight begins on the beach at Puerto Vallarta in front of the Hotel Fiesta Americana.

above 20 knots. Takeoffs and landings are made from the sandy beach. We were tempted but didn't try the parasail (Figs. 10-37 through 10-39).

Much has been written about the history of Puerto Vallarta, and you'll find all the detail you want in any pre-flight planning book. Capsulated, here's how *Guide* magazine presented the history of the town.

"Puerto Vallarta was discovered by Don Pedro de Alvarado in 1541 some 20 years after Hernan Cortes had conquered the Aztecs. From that point on it was forgotten; rumor has it pirates used the port, but for what, is not known. In 1850 Don Guadalupe Sanchez moved his family to the area and settled along the Caule River. During that period it was called Puerto Las Peñas. It was not until 1918 that the area was called Puerto Vallarta and has its first mayor, Jesus Langarica. The name Vallarta was in honor of the governor of Jalisco, Don Ignacio Luis Vallarta.

"Over the years the fact Puerto Vallarta has been hard to get to is the reason why today, rather than being a large plastic resort, the town still has the characteristics of a quaint Mexican village with cobblestone streets, red tiled roofs, bougainvillas, and blue jacaranda trees galore (Fig. 10-40).

"When John Huston decided to make *The Night of the Iguana* in 1963, Puerto Vallarta came into its own. With Elizabeth Taylor and Richard Burton's personal and romantic interest in the town, Puerto Vallarta fast became a resort. Jet airline service and a new airport were soon to follow and by 1970 a new fully paved highway from Tepic connecting Puerto Vallarta with the outside world was in full use. Progress is changing Puerto Vallarta, or simply "PV" as it is referred to by many Americans, but

Fig. 10-39. Parasail gains altitude at Puerto Vallarta. Operator of this thrill ride, Alfredo Almaras, "test hops" his equipment each day before passengers are flown.

Fig. 10-40. Downtown Puerto Vallarta along the shoreline. Broad sidewalks border Avenida Diaz Ordaz. Hotel in the background is the Oceano.

somehow it has managed to maintain its charm and therein lies the secret (Figs. 10-41, 10-42).

"Yelapa: The daily boat excursions to Yelapa are a must in Puerto Vallarta. With its Gauguin setting, thatched roof native houses, curving

Fig. 10-41. Seahorse and mermaid statue is one of the most photographed statues in the area.

Fig. 10-42. Aerial photo of the Zocalo area of Puerto Vallarta. Seahorse statue is along the shoreline in the middle of the photo. Oceano Hotel, center, overlooks the *Malecon* ("boardwalk").

white sand beach, lagoon, hotel, seafood restaurants and shops, Yelapa is a paradise for tourists who want to get off the beaten track and visit a primitive, quiet beach reached only by sea. The yachts *Sombrero* and *Palladin* make daily trips, leaving the Harbour at 9 a.m. and the main pier (Los Mochis pier) at 9:30 a.m., returning to P.V. by 4:30 p.m. A beach bay area is inhabited by some loveable but slightly crazy Americans who have found Yelapa the fulfillment of the get-away-from-it-all dream. No roads or electricity in this picturesque setting of a Pacific island paradise. The Hotel Lagunita Yelapa has thirteen Palapa thatched cottages in a Tahiti setting. Rita Tillett's handcrafts and Indian clothing Bazaar and Native Home is a must visit.

"Isla Del Rio Cuale: The town of Puerto Vallarta is divided by the River Cuale with two bridges that cross it. In the middle of the river is an island that in recent times has been developed into a unique open-air shopping mall along the part of the island closest to the ocean. There are shops, restaurants and even a small museum. Just the walk is worthwhile. Further up the river there is a children's playground, an art school and even open-air art exhibits. And despite this development, you can still see the Indian women out washing clothes in the river as they have done for centuries. Further up the river is the area called Gringo Gulch where originally most Americans built their homes, including Elizabeth Taylor and Richard Burton."

Puerto Vallarta is a delightful place to visit. We stayed a day longer than originally planned and could have spent considerably more time in the area. However, there was much more of the West Coast to rediscover.

Chapter 11

West Coast Beyond Puerto Vallarta

When you look at a map of Mexico, Puerto Vallarta is as good a place to split the country as Guadalajara or Mexico City. When you fly the west coast, there's little or nothing (in 1980 construction) between Vallarta and Acapulco except for Manzanillo and then on to Zihuatanejo and Ixtapec. However, this lush, almost forgotten coastline is certainly destined for an exploding development both in tourism and in resort areas for Mexicans escaping the smog of their capitol city. It's a great yet unspoiled coastline as we write about it. Tomorrow may be a different story.

Departing Vallarta (Fig. 11-1), we looked for the movie location of the film *Night of the Iguana*, but we weren't sure we had found it. Again, we cruised down the coastline and overflew the way-out-of-town airport at Manzanillo. We talked with the tower to assure that 100-octane was available for a stop on our return trip. All the ASA fields we visited had an ample supply of av/gas and most of the pumps had both 80 and 100 octane (Fig. 11-2).

MANZANILLO

When we did land at Manzanillo later in the trip, we took on a load of fuel and had lunch in the beautiful terminal building, but did not take the 40 kilometer, 420-peso cab ride to town. At the airport, a Denver-based Boeing 707 Ports o' Call tour aircraft was waiting for a load of tourists to complete their sun-and-fun activities and return to the U.S.

Manzanillo is considered Mexico's prime Pacific port (Figs. 11-3 through 11-6). The town itself is geared toward shipping with a bronze bell atop one of the nearby hills to announce approaching vessels. Much of the recent development is northwest of town along Mexican Highway No. 200.

Fig. 11-1. Downtown Puerto Vallarta looking back toward the new high-rise resort buildings and the airport.

Las Hadas, a 203-unit four-star motor hotel, is located 11.5 km up this road and 2.5 km on the Peninsula Santiago. Arabesque architecture, patios, pools, and balconies are outstanding with private pools in suites. There are eight tennis courts, boat dock, miniature golf courses, and predictably high prices. Some wags have dubbed the structure "Disneyland South," but it is a striking assembly of buildings (Fig. 11-7). Two additional large hotel/ motel complexes—the Hotel Playa de Santiago and Vida del Mar Condominiums—are farther up the shore to the west.

Fig. 11-2. Airport terminal complex at Manzanillo. ASA service was efficient and credit cards were okay for fuel. Here's an airport that is truly right on the beach.

Fig. 11-3. Manzanillo is an excellent seaport for Mexico. Deep water ships can load directly from these wharfs.

All indications point to a continued growth of large hotel, motel, and condo developments in this tropical area with excellent fishing, skin diving, water skiing, and surfing, particularly at Cuyutlan, 42 km southeast of Manzanillo where the "green wave," 15 meters high, is found.

Fig. 11-4. The downtown area of Manzanillo is quite small in area. Ships anchor in the harbor awaiting docking space.

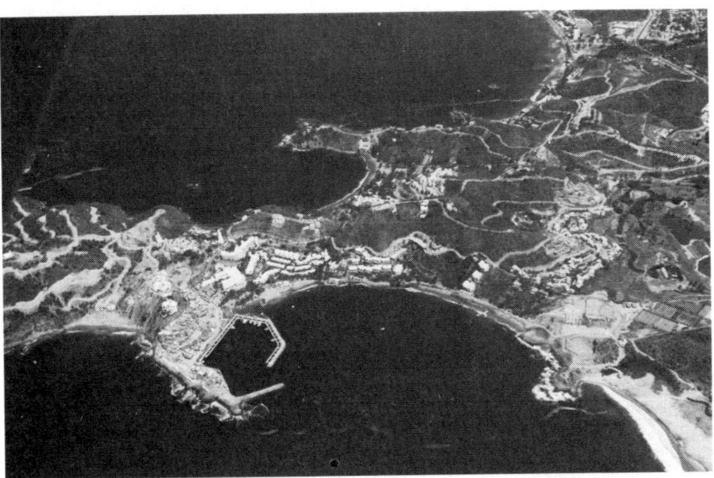

Fig. 11-5. High-altitude aerial photo of the Las Hadas area. New development continues surrounding the present resort (courtesy Mexican National Tourist Council).

All along this south coast there were large and small resort areas in various stages of development. With banana and coconut plantations along the shoreline and heavy tropical jungle inland, the area is truly a tropical paradise (Fig. 11-8).

When flying this portion of the Mexican west coast, remember that your course is essentially west-to-east. Heading down the coastline here,

Fig. 11-6. Like something out of "Arabian Nights," Las Hadas is a striking example of new luxury living along Mexico's tropical southwest coast (courtesy Mexican National Tourist Council).

Fig. 11-7. A closer view of the architecture at Las Hadas shows the intricate structures of "Disneyland South." Note new construction at the left of this photo. The area was opened originally in 1974 by Bolivian tin magnate Antenor Patino. The original resort had 200 rooms on 12 acres covered with 3000 palm trees (courtesy Mexican Tourist Council).

you can easily lose an hour of sun time in a half-day's flight, so plan accordingly. Even today, lighting at all but Mexico's key airports can leave something to be desired.

Southeast of Manzanillo (heading 109°), the WAC chart was completely out-of-date. The road listed as "under construction" is nearing completion and is much closer to the shoreline than the depiction (Fig. 11-9). At least two paved airports are not shown on CJ-24 (Fig. 11-10). A single pilot attempting to navigate with this chart in restricted visibility could have all kinds of problems. However, you can stay offshore and

Fig. 11-8. Small resort town south of Manzanillo. Note thatched roofs of pagodas for sunbathers and coconut plantation at the top of the photo. This photo was taken fairly early in the day before sunbathers had arrived.

eventually pick up the Zihuatanejo VOR. The large open-pit mines and a shoreline steel mill at Lazaro Cardenas, 50 miles northwest of Zihuatanejo at the beginning of Bahia de Petacalco, are not depicted on this chart at all and can lead you to doubt your location. Our notes show the comment, "Love those long-range tanks," that was not transmitted in flight over our four-way intercom. We told our passengers about it once we were on the ground. And that extra fuel on board was comforting when we couldn't pinpoint our location with certainty.

Fig. 11-9. Highway construction is time-consuming and difficult along the rugged coastline between Manzanillo and Zihuatanejo. Eventually this will be a scenic, curving tourist highway.

Fig. 11-10. This paved airport beside the new road still under construction was not marked on the latest CJ-24 charts. From all appearances, the strip was put in to facilitate road building.

ZIHUATANEJO

Zihuatanejo is a small town with perhaps 2000 residents and was once an important shipping port for bananas. The town retains the leisurely pace of the tropics and, for us, was a delightful change of pace (Fig. 11-11, 11-12). Nine kilometers to the north is Ixtapa where new high-rise tourist

Fig. 11-11. Aerial photos of "downtown" Zihuatanejo. Some 2000 people live here year-round (courtesy Mexican Tourist Council).

Fig. 11-12. Small fishing boats are beached under palm trees in the downtown area of Zihuatanejo.

hotels are blossoming. We chose another one-star unit, the Hotel Irma, near "downtown" Zihuatanejo, and saved another bundle which we promptly spent on local clams, shrimp cocktails and margueritas.

Once there was a good dirt airport almost in town. The new jetport, with one of the nicest terminals we've ever seen, was 18 kilometers and 50 pesos each (fixed transportation charge) out of town (Fig. 11-13). In con-

Fig. 11-13. Jet-sized International Airport at Zihuatanejo is a long cab ride from town. The facility is excellent but at present is seldom used except for long-range jet transports.

Fig. 11-14. New hotels on the beach at Ixtapa in 1980 along the Playa Vista Hermosa. In the foreground is the Hotel Viva followed by the Holiday Inn, Riviera del Sol, Presidente and Aristos (courtesy Mexican Tourist Council).

trast with the usually abundant helpful airport attendants, there was no effort to handle baggage or assist in any way at this terminal. Perhaps there is so little private flying here that the airport isn't geared for tourists in small airplanes.

As in other parts of the world, there seems to be a universal macho among cab drivers who take a fiendish delight in proving that they have more nerve than any pilot as they dodge pigs, cattle, and children on the highway. It's no contest.

Zihuatanejo was certainly the most colorful, fully tropical, yet unspoiled town on our itinerary. Shops are filled with more-or-less authentic mementos. Sidewalk cafes are friendly and busy late in the afternoon. Side streets are incredibly rough and the cab ride to the hillside location of Hotel Irma was rugged for both cab and passengers. The old town itself is as close to unspoiled Mexico as we found. It is certainly worth visiting again, but maybe during the cooler season of the year, although once again the Casablanca fans were excellent.

Zihuatanejo is a sheltered port that once challenged Acapulco for the China trade. It was chosen by Mexican resort developers for some major investments and, under Government sponsorship, a whole new area at Ixtapa, about nine kilometers northwest, is being developed rapidly (Fig. 11-14). When we flew over the area in mid-1980, a half dozen new resort hotels, swimming pools, and a golf course were underway. Sponsored by *Fonatur* (Fondo Nacional de Fomento al Turismo), the 15-mile stretch of palm-lined, blue-water beaches with bays and coves, rich vegetation, and a series of close-in islands is sure to become a tropical resort area that may someday rival Acapulco.

The harbor at Zihuatanejo has four main beaches: downtown, Madera, La Ropa and Los Gatos, the most beautiful in Zihuatanejo. It is said that the Tarascan chief who used the beach as his private recreation area was afraid of being attacked by wildcats infesting the surrounding jungle. He built a system of barriers and traps to keep the marauders at bay, thus the name *Los Gatos*, meaning "The Cats."

Of all Zihuatanejo's beaches, Los Gatos is the most Polynesian-like, featuring an offshore coral reef as well as a grove of sheltering palms, clear blue water, and soft, fine-grained sand. It is also the least accessible and can only be reached by boat. The trip is well worth making if just for the atmosphere. Los Gatos has several thatched-roof, open-air restaurants featuring such marine delicacies as lobster, clams and oysters fresh from the ocean, and charcoal-broiled red snapper and roosterfish; scuba diving and snorkeling equipment is available to play around in the clear waters with instructors and guides to assist.

We spent an extra night in Zihuatanejo, enjoying the balmy climate where the average dawn-to-dusk temperature is 80°F for over 200 days a year. Even without visiting the new tall towers of Ixtapa, you can find a great variety of small shops featuring jade, tropical clothing and even a curio shop featuring goods from Damascus and Bagdad that's called the "Cashbah." Local seafood is just great. Red-shelled clams are a specialty of the area.

Fig. 11-15. Hotel row at Ixtapa looking northwest. In the foreground is the Hotel Aristos followed by the Presidente, Riviera del Sol, Holiday Inn and Hotel Viva (courtesy Mexican Tourist Council).

Fig. 11-16. New high-rise luxury hotel nears completion at Ixtapa. Development here has been sponsored by the Government agencies FONATUR (Fondo National de Fomento al Turismo) and FIBAZI (Fideicomiso Bahia de Zihuatanejo) (courtesy Mexican Tourist Council).

Ixtapa is a new area isolated from Zihuatanejo by a low ridge of hills (Fig. 11-15). A new 18-hole, 6910-yard golf course designed by Robert Trent Jones opened in 1976 with a residential area surrounding the greens. A large modern shopping center adjoins the golf course and caters to guests of the new high rise hotels, El Presidente, Las Palmas, Aristos Ixtapa and the nearly-completed Holiday Inn, Condominium Playa Sol, and Hotel Famitur (Fig. 11-16). Playa Quieta, 10 kilometers west of the golf course, has crystal clear water, reknowned vegetation, and even a diving school called Oliverio. Excellent diving sites are nearby and on the Island of Ixtapa, a wildlife preservation park where hundreds of wild birds and animals can be seen and photographed. Animals here include deer, badger, armadillo, racoon, iguana, parrots, and seagulls.

This general area is a tough one to leave, but leave we did.

ACAPULCO

When we departed for the one-hour short hop to Acapulco, the tower said, "Good day and good luck" (Fig. 11-17). We watched fisherman near the shore and shrimpers far offshore before coming in over the military airport at Pie de la Questa (Figs. 11-18, 11-19). Many years ago, this was the main airport for Acapulco, a truly tropical setting with the runway cut into a coconut palm grove. Today, a couple of military helicopters can be seen, but there appears to be little or no action.

As you contact the tower and fly in over Acapulco's beautiful harbor (Fig. 11-20), you know that you're back in civilization, along with a quarter-million residents plus hoards of sun-loving tourists. White high-

Fig. 11-17. Tower operator at Zihuatanejo works with modern communications equipment. Mexican tower operators speak acceptable-to-good English in almost all cases. Note the red/green/white "biscuit gun" for no-radio aircraft hanging above the operator's head.

rise hotels and condominiums vie for space along the bay. Sparkling cruise ships dot the harbor (Fig. 11-21). It's a big, bustling city that offers just about anything, and that means *anything* that you might want to see or do (Fig. 11-22).

Fig. 11-18. Military airport at Pie de la Questa northwest of Acapulco was once the resort's main airport. It is now used only by a few military aircraft. Don't land here!

Fig. 11-19. Palm trees and wide beaches lie beside the picturesque Pie de la Questa Airport. Note wide lagoon in the background.

The airport is big and the general aviation terminal (called the Executive Terminal) is sparkling clean (Figs. 11-23, 11-24). Service is courteous, and rapid by Mexican standards. The parking areas held a couple of small jets, one or two heavy twins and a handful of four-placers. The main passenger terminal is under expansion and just as hectic as any other international jet terminal. We enjoyed a good lunch (try their *chilquiles*) at the Wings Cafe on the second floor of the terminal. Their ice cream is good, too.

The most economical way into town from the airport is by an airport "limo." These cars or minibuses will carry five passengers at a fixed rate

Fig. 11-20. Acapulco Bay is almost a completely enclosed harbor. The airport is beyond the low ridge in the background.

Fig. 11-21. Looking down the Costera M. Aleman, one can see almost wall-to-wall high-rise resort buildings. From foreground to background are the Paraiso Marriott, Ritz, Continental, Torre de Acapulco, Diana Fountain, Condesa del Mar, El Presidente, El Matador, Holiday Inn and Acapulco Malibu Beach hotels (courtesy Mexican Tourist Council).

about half that of a taxi. The cabs in Acapulco don't have meters, so you must make your own deal before entering. Hertz and Avis have rental agencies at the airport if you plan to do a great deal of driving and choose to accept the problems of driving a strange car in a very strange city.

Acapulco is . . . well, it's just Acapulco. There's nothing else in Mexico even approaching it. Take the quarter of a million permanent residents plus more than 3-½ million visitors annually, and you can rightfully expect an international town that has just about anything to offer that you can think of, and more.

Like the remainder of Mexico's tropical west coast (actually Acapulco is almost due south of Mexico City with the airway radial 006°), the main

Fig. 11-22. Acapulco Harbor as seen from the air with a string of new, posh, high-rise hotels along the beach (courtesy Mexican Tourist Council).

drawing card is the weather. Of course, you can visit Acapulco and do nothing but lie on the beach to enjoy the sun and the people, but chances are good that you'll do the town. And that's an undertaking in selection. Hotels are expensive—twice as expensive from mid-December through April with rates of $100 per night for a double room. There are now over 350 hotels here (Figs. 11-25 through 11-29).

Fig. 11-23. Acapulco Airport as seen on downwind leg for Runway 28. Jet aircraft are parked away from the terminal and passengers are transported by special buses.

Fig. 11-24. Modern general aviation terminal building at Acapulco. Facilities here are much more comfortable than the jammed, expanding airline terminal building.

Veteran international publicist Bud Lewis describes some of the luxury hotels in these words:

"Acapulco is the home of the famed Las Brisas, an elegant colony of casitas dotting the hillside over the bay, 250 private villas, 200 with their own private swimming pools. Each casita has its own marble bath, refrigerator and terrace. The larder is stocked daily with fresh fruit, bottled water and Mexican beer. The color for Las Brisas is pink, and everything that will take paint bears that color. Pink swimming pools are strewn with pink hibiscus blossoms. A call from your pink telephone brings around a pink jeep, with or without driver (not pink).

"The Acapulco Princess Hotel started with the basic design of an Aztec pyramid, but from that point on forgot all about the Indians and went directly into the future. A live lagoon bubbles through five small groves of 60-foot palm trees. *La Gente Bonita* (even in Spanish they have a name for the 'Beautiful People') sit in string bikinis, gently tasting their Tequila Sunrises at the Lagoon Bar. Gazing upwards to the 16-story-high ceiling is a visual treat as cascades of tropical flowers tumble from tiled balconies. And this is just what you see on a simple stroll through the lobby.

"The Princess, with 777 rooms, sits on Revolcadero Beach, just south of the city, surrounded by 200 acres of championship golf course, tennis courts, residential villas, a huge pool featuring an under-the-waterfall swim-up bar.

"Hotels range from the spectacular to the budget special. The 800-room Plaza Internacional built for $36 million, has a bedroom swimming pool in the 23rd floor Presidential Suite. The plush 500 room Condesa Del Mar, operated by American Airlines, is a top choice in both winter and summer seasons.

"The Villa Vera features cottages high in the hills overlooking the bay. Ten swimming pools here—nine of them private. But the big feature is a bar in the pool, complete with underwater stools. The beautiful Acapulco Continental midway on the beach facing the bay features a gigantic pool that looks more like a moat. Swimming in a grand circle, you can stop at the bar or grab a bite without ever leaving the water. When you're ready for a second round, you can justify it by swimming another lap."

"A late entry in the total luxury resort class is the La Palapa, right on the beach, soaring 30 stories. Centered amid lush gardens and winding

Fig. 11-25. Aerial close-up photo of a typical Acapulco high-rise hotel some 20 stories high where every room has a balcony that overlooks the bay. Note thatched umbrellas along the beach (courtesy Mexican Tourist Council).

Fig. 11-26. Hotels at Acapulco are jammed tightly together along the shorefront. Note how quickly the land becomes jungle in the background (courtesy Mexican Tourist Council).

pathways, La Palapa features large swimming pools, beautiful beach frontage, and a complement of colorful restaurants, bar and shops. Each of its 400 suites (there are no just rooms) has a private terrace overlooking the ocean."

Everyone has their own favorite spot in Mexico. With George and Madge Craig from Milpitas, California, it's the Hotel Mision on Felipe Valle in Acapulco, which is not even listed in the AAA travel guide. We'll let them tell it in their own words and also tell about their choice in Zihuatanejo:"

"There are little pensions on the back streets of Acapulco where the prices are reasonable and where the international tourists congregate (Fig. 11-30). One in particular, the Hotel Mision managed by the Senora Melena on Felipe Valle in Acapulco, used to be a grammar school. In fact, the street in front of the hotel, Felipe Valle, is named for the original school master who ran the school in what is now the hotel. Across the street from the hotel is a kindergarten school. The hotel is made up of rooms around a central patio. There is a gate that is locked at night that opens onto the street.

Across the patio from the gate, there are three stories of rooms. Air conditioning is provided by Casablanca fans which are surprisingly effective. There is no glass on the windows—only screen wire—as there is no need for windows in the climate of Acapulco. The guests help run the hotel. When they are in the patio, they will substitute at the desk for the clerk who may be out doing something else. The atmosphere is astonishingly friendly and relaxed. It's the kind of informal accommodations that one can find in Acapulco in contrast to the chrome and glass skyscrapers one usually thinks of when Acapulco comes to mind.

"Another delightful hotel, Sota Vento in Zihuatanejo, is one of the older of the newer hotels. It is built on the side of the hill leading down to Zihuatanejo bay and a set of rooms has a terrace that forms the roof of the rooms below, providing each with a view through the tropical foliage of the bay. In the evening with the tropical sunsets, it's a particularly enchanting hotel and quite reasonable. When we were last there, the price during the tourist season was $40 for two with two meals per day."

When it's time to eat, Acapulco has everything from Kentucky Fried Chicken, Big Boy, or Pizza Hut right on up the line to some of the finest

Fig. 11-27. Sunbathers loll on the beach at Acapulco as a vendor attempts to sell clothing (courtesy Bud Lewis, The Lewis Co., Ltd.).

Fig. 11-28. Acapulco Princess Hotel and golf course are located near the International Airport. Basic design is of an Aztec pyramid (courtesy Mexican Tourist Council).

restaurants in the world. Although Acapulco features an international cuisine, there are American favorites prepared Mexican style. There's nothing quite like a *hamburguesa* followed up with a *maltesa. Ceviche,* the native dish of Acapulco, is a must. It's raw fish seasoned with lime and

Fig. 11-29. The 800-room Plaza Internacional Hotel is located at the far east end of Acapulco Bay. Note private balconies facing the ocean (courtesy Mexican National Tourist Council).

Fig. 11-30. Town square and church in the older section of Acapulco where smaller fishing boats can be chartered. Many small hotels in this area are less expensive than the beach-front high-rise units.

spices, and there are as many variations of it as there are fish in the bay. It's meant to be washed down with a couple of cold *cervezas*.

A number of attractions are unique to Acapulco. Perhaps the best known are the Acapulco Divers at La Quebrada where dives are scheduled every evening at 8:15, 9:15, 10:30 and 11:30 (Fig. 11-31). Over twenty-five years ago the young men of Acapulco used to dive from various levels of La Quebrada cliff for their own amusement or for a few coins given to them by tourists. One day, Teddy Stauffer, a Swiss who came to Acapulco with Errol Flynn and never left (he is now owner of Sunset Restaurant Bar & Beach Club) was watching them and conceived the idea of building a restaurant down the cliff, below the Hotel El Mirador, so that the patrons could enjoy the spectacle of the 135-foot dives. The restaurant became very popular, and the divers internationally famous. One can also watch the dives from a peninsula across the cliff for a small entrance fee. There are no scheduled dives in the daytime. Frequently, however, the city tour, which can be organized through the travel agent in your hotel, arranges a dive. Many people just wander over to La Quebrada and are fortunate enough to find the divers practicing.

Acapulco's Convention Center, completed in 1973, has large and varied displays. The auditorium seats 5000 and features big name entertainers. The Plaza Mexicana presents a Mexican fair twice nightly featuring the Acapulco High Divers, Flying Indians, fireworks, Ballet Azteca, Mexi-

Fig. 11-31. Low-altitude aerial photo of La Quebrada where the divers plunge 135 feet into the ocean. Arrow points out the diving platform. Hotel El Mirador is in the background (courtesy Mexican Tourist Council).

Fig. 11-32. Both white cruise ships and cargo container vessels tie up at the Acapulco Harbor. Excursions around the harbor can be taken from this general area (courtesy Mexican Tourist Council).

can Dances and a long string of entertainers including the Mérida Folkloric Ballet. The same Convention Center complex houses a 1500-seat motion picture theater and a 2500-seat open-air amphitheater, the night club El Internacional with Las Vegas-type extravaganzas, El Chef restaurant featuring international and Japanese cuisine, the Disco Laser, and the Tablado Andaluz featuring Spanish Flamenco art.

In the way of sports (not sporting), you'll find three golf courses, a variety of tennis courts, two skin diving schools with boats and all equipment, sport fishing boats, parasailing, and boat trips around the harbor both day and night (Fig. 11-32). One cruise boat, the *Bonanza,* departs daily at 10:30 a.m. with a platform that lowers to sea level for swimming. There's a snack bar, orchestra, and an air-conditioned section and drinks are included in the price of the ticket. Time en route, including a half-hour stop at Majagua Beach in Puerto Marques, is 2½ hours.

If you want to learn to water ski, *Guide* magazine recommends *Escuela de Skiis.* Glorieta Tlacapanocha speaks perfect English, has taught countless celebrities and is cautious and reliable. You can trust your children or your grandmother with them."

ON TO TAPACHULA

Enjoy your stay in Acapulco. When you run out of time or money, or both, then you can continue on down the west coast of Mexico that swings northeast beyond Puerto Angel. Today there's a secondary road that stays with the beach, but when we flew it many years ago in a Navion, this was a most lonesome stretch—reputedly inhabited by headhunters.

Our temperature gauge read an even 100°F on takeoff from Acapulco but by the time we climbed to 2500 feet things were down to a tolerable 82°. It isn't as scenic, but certainly more comfortable to cruise at a somewhat higher altitude during the middle of the day when the temps are really high (Fig. 11-33).

You'll soon pass from the State of Guerro to Oaxaca. Your charts will show many small dirt flight strips; most are very short, almost all are unimproved, and many are marked private. The first paved airport is the new 5200-foot flight strip at Puerto Escondido. Puerto Escondido is at the end of a poor highway and reportedly has a charm all its own. Bill Robinson (Vice President, Communications, Beech Aircraft) thinks highly enough of this hideaway resort to spend his Christmas here. When asked what there was to see and do, he merely smiled without comment. If you're flying down this coastline Puerto Escondido might just be a delightful grabbag. We've flown over it but never landed.

South of Puerto Escondido, you'll have to shuffle charts since CJ-24 stops here and CJ-25 begins. The next city of any size is Tehuantepec, once the stronghold of the original inhabitants of the region. One travel guide says that it is celebrated for its fiestas, hot springs, and handsome women, the Tehuanas. The main airport here is marked military and should not be used without prior permission. A smaller dirt strip on the coast at Salina Cruz is much closer to town, but its condition is not known. Proceed with caution. The latest AAA Guide doesn't list either the town nor any hotels.

To continue east to the Guatamalan border, you'll need chart CK-25 or

Fig. 11-33. New highway construction along the coastline in the southwest area of Mexico. The majority of tourist travel is by air or ship and will remain that way until these roads have been completed.

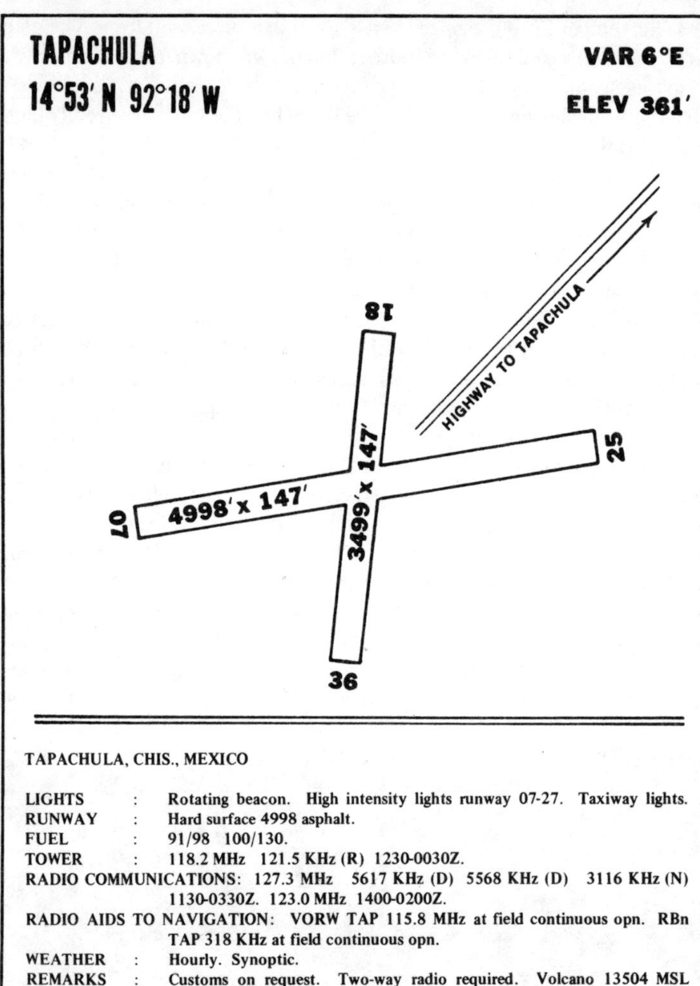

TAPACHULA, CHIS., MEXICO

LIGHTS	:	Rotating beacon. High intensity lights runway 07-27. Taxiway lights.
RUNWAY	:	Hard surface 4998 asphalt.
FUEL	:	91/98 100/130.
TOWER	:	118.2 MHz 121.5 KHz (R) 1230-0030Z.

RADIO COMMUNICATIONS: 127.3 MHz 5617 KHz (D) 5568 KHz (D) 3116 KHz (N) 1130-0330Z. 123.0 MHz 1400-0200Z.

RADIO AIDS TO NAVIGATION: VORW TAP 115.8 MHz at field continuous opn. RBn TAP 318 KHz at field continuous opn.

WEATHER	:	Hourly. Synoptic.
REMARKS	:	Customs on request. Two-way radio required. Volcano 13504 MSL 18.6 NM magnetic heading of 030 from airport.

Fig. 11-34. Sketch of the layout of the airport at Tapachula, Mexico, just north of the Guatamalan border.

a reasonable facsimile. The Pan American Highway stays fairly close to the shoreline while a good paved highway swings inland through Tuxtla Gutierrez and San Cristobal de las Casas. Both of these towns have airports with some services. Tuxtla Gutierrez has two good-sized hotels—the Bonampak and Gran Hotel Humberto—while San Cristobal de las Casas lists only two—El Molino de la Alboriada adjoining the flight strip and the 36-unit Hotel Espanol.

End of the line on the Mexican southwest coast in Tapachula (Fig.

11-34) just a dozen miles from the Guatamalan border. On a visit here with a Cessna 207 on a delivery flight, we called the tower on a blustery Sunday afternoon and received no reply. The operator was on a siesta. However, there was some activity on the airport and we had to clear Customs anyhow so we landed. Fuel was almost not available because of local agricultural aircraft needs to combat hordes of insects. However, we were able to fill one tank and reach Guatamala City the next day with no problem other than paperwork.

The town is clean and comfortable with shaded tropical trees and an attractive plaza. The town has two large motels, the Kamico and Loma Real, and the downtown Hotel San Francisco. The hotel was comfortable and the food good. However, Tapachula is not truly a tourist resort area and is used primarily by pilots heading deep into Central America.

Chapter 12
Baja California

If you live in California (as we do), your first choice for an international flight is usually Baja California. We've been doing just that for the past 20 years. It all began when we took a Cessna 172, N2058Y, from Long Beach to the tip of the Baja Peninsula on a magazine assignment (Fig. 12-1). A year later, we took two Pipers and produced an interesting half-hour color and sound travelogue, *Wings to Baja*, for Piper. One of the early 160-hp Cherokees was our touring airplane (Fig. 12-2), and we shot pictures out of a 250-hp Comanche. The peninsula of Baja has long been a favorite piece of land for us.

Actually, there are two faces of Baja. Before the paved transpeninsular highway was completed in 1973, Baja was an unspoiled hideaway enjoyed primarily by yachtsmen, rugged four-wheel drive operators, and lightplane pilots (Fig. 12-3). Came the highway and you now stand in line with your airplane while station wagons, campers, and motorcycles line up at the av/gas pumps at some of the smaller resort airports.

However, progress has some virtues. The influx of tourists has created better hotels, better water systems, more stores and fishing launches. On an early trip to Bajia de Las Angeles (Fig. 12-4) a very tired young man driving a flatbed truck asked for "anything" to stop his diarrhea. While it wasn't professional, we gave him a couple of Lomotil tablets that did the job. The next morning he thanked us and asked if we could telephone his sister in Los Angeles on our return, advising that he was in good shape and had gone as far as L. A. Bay. Without thinking, we said, "Sure, we'll call her sometime late this afternoon." His reply was, "You mean that you're going to be back in the States today? I've been driving for a week over some of the roughest roads I've ever seen just to get this far."

Fig. 12-1. Cessna 172 on an early flight into Baja. The "airport" is at Bahia de Palmas, midway between La Paz and San Jose del Cabo.

Fortunately, there are still some isolated beaches and villages not yet touched by "the road." They remain much the same as the area was decades ago. We'll mention those that we know of personally as this report continues.

TIJUANA

The two international airports of entry for Baja are Tijuana and Mexicali. Because of the volume of traffic through these ports, the procedure is

Fig. 12-2. Piper Cherokee used in filming *Wings to Baja* in the early 1960s is shown in flight near Hotel Cabo San Lucas.

Fig. 12-3. Early spacecraft photo of the Baja Peninsula looking down the west coast. Large bay in the middle of this photo is Bahia de Sebastian Vixcaino at the 28th parallel that separates the states of Northern and Southern Baja (courtesy NASA).

fairly well streamlined and paperwork is kept to a minimum. While both border towns are popular for tourists with automobiles, they're merely the jumping-off spot for lightplane pilots. You do your paperwork, top your fuel tanks with the good economy PEMEX, and head south.

Fig. 12-4. Refueling at L. A. Bay. Pioneer store established by Antero Diaz had adjoining inexpensive cold-water bungalows for visitors. Aircraft landed here right in the "front yard" of the resort. Recently, a large new paved airport has been built a mile away.

Fig. 12-5. Mexicali, Baja California Airport parking ramp with control tower and all border-crossing facilities.

Mexicali as a port of entry was covered in the chapter on the West Coast (Figs. 12-5, 12-6). Tijuana really doesn't differ except in the volume of traffic of Aero Mexico where you can see two or three of their DC-9s on the ramp at one time. The passcnger terminal is spacious and has all the amenities that the fare-paying passenger requires, but the lightplane pilot

Fig. 12-6. PC-7 two-passenger turboprop operated by the Mexican Air Force is parked on the ramp at the Tijuana International Airport. Complete border-crossing facilities are available here and the new fee of 150 Pesos for VFR and 400 Pesos for IFR flight plans is now charged here.

should have no reason to visit that side of the field. Fuel, flight planning, tourist cards, and all the essentials are available right at the base of the tower. The only thing you need that you can't get at the airport as this book is prepared is Mexican liability insurance, so be sure to have that in advance as is noted in the pre-flight chapter.

ENSENADA

Out of Tijuana and down the west coast, your first stop can be Ensenada, a mere 40 miles south of the border. As this book is being prepared, you can fly or drive as far as Ensenada without a tourist card. Ensenada, for us, is a great place to go shopping. Prices are as low as anywhere in Mexico—far lower than Mexico City—and you can buy just about anything you want along Lopez Mateos or Lazaro Cardenas streets, just inland from the Bahia de Todos Santos. Many Californians, the authors included, will pay the $5 cab fare each way from Ensenada's Airport to town just to go shopping and get a good seafood dinner. There are fine motels for those who wish to spend the night, have a fine inexpensive dinner and sample the local liquors (Fig. 12-7).

The airport in Ensenada is "joint use" with a fairly large contingent of Mexican military personnel on the field and a guard at the gate. Chances of having your aircraft molested here are minimal (Fig. 12-8).

Like so many towns in Mexico, the origin of Ensenada was a ranch given by royal grant to a Spaniard for services rendered the crown. Ensenada, which means inlet or cove, is well-named when you think of its beautiful bay (Fig. 12-9). It was once a gold boom town, served as capital of

Fig. 12-7. Airport at Ensenada is a short cab ride south of town. This aerial photo was taken during the start of one of the Baja road races and shows an unusually large number of aircraft on the ground. The taxiways have since been paved.

Fig. 12-8. Small but adequate passenger terminal and control tower cab at the Ensenada Airport. Fuel is usually available, but credit cards are not accepted here.

the territory of Baja California Norte for a short period, and enjoyed a second boom when gambling was legal back in the '30s. It is also a fishing center, especially for the sportsman. Aside from being a favorite resort for Californians, it is well known as the Yellowtail Capital of the world. Its annual events include the Newport Beach to Ensenada Yacht race, Tecate to Ensenada Bicycle race, and the Baja 500 and 1000.

Fig. 12-9. Harbor of Ensenada handles many cargos. White "mini" cruise ship is a frequent visitor. Note cargo of cotton awaiting shipment.

Most visitors flying into Baja will continue deep into the peninsula rather than choose a resort so close to the U.S. border as to be almost "stateside." However, two old-time fly-in resorts are located within about 150 miles of the border, east of the towns of Colonet and Camalu along a winding dirt road. Meling Ranch has a 4000-foot graded airstrip, elevation 2280 feet. The area has a 10,000-acre cattle ranch, offers horseback riding, good accommodations, and family-style meals. For reservations or just information write Apartado Postal (P. O. Box) 224, Ensenada, Baja California, Mexico.

Farther up this same road is Mike's Sky Ranch, complete with a 4750-foot flight strip at an elevation of some 4500 feet. Visitors here can hunt deer, cougar, bobcat, quail, or dove. For information or reservations write 1000 Avenida Revolucion, Tijuana, B.C., Mexico. Mike's Sky Ranch is just on the eastern border of the Parque Nacional Sierra San Pedro Martir, where a modern astronomical observatory was built by the Mexican government near the 10,126-foot Picacho del Diablo.

Along the West Coast, the paved road continues along the shoreline as far as El Rosario before turning inland (Fig. 12-10). On the eastern side, there's a paved road over desert and salt flats to San Felipe where there's a fine paved airport and little activity. A fair dirt road goes as far south as Puertecitos where you'll find an adequate dirt flight strip right in the middle of the village.

The peninsula narrows in this area and pilots can see both sides easily from 4000 or 5000 feet. Most dependable of the fueling stops in northern Baja is at Santa Ines where av/gas, pumped through two filters, is nearly always available. The runway is paved, there's a small cafe and austere sleeping rooms adjoining the airport, and an El Presidente Hotel within a mile at Catavina. The area is striking in its rock formations and sparse desert shrubs (Fig. 12-11).

Fig. 12-10. Airport at El Rosario with race support aircraft on the ground. This airport, as well as the road, has since been paved.

Fig. 12-11. Towering cactus grow in the area near the Presidente Hotel at Santa Inés. This is truly an oasis in a dry desert.

Almost every fishing village on the peninsula has a flight strip of sorts. Arnold Senterfitt's Baja Bush Pilots take a great delight in exploring these truly isolated areas. We'd recommend this type of exploring in a true STOL aircraft, preferably in a flight of two, and with high-time pilots who can land on-the-numbers every time. Otherwise, the wilds of Baja are a very poor place to bend up an airplane.

A number of such strips are scattered around Bahia San Luis Gonzaga on the Gulf side just south of Santa Inés. We've landed at a couple of them, Alfonsina's and Punta Final. Depending on the time of year and when the last rain fell, the strips are either good or not-so-good. Punta Final, site of a small resort that didn't survive, has one of the most beautiful white sandy beaches that you'll ever see. We spent a night there a number of years ago while flying a venerable Piper PA-12.

BAHIA DE LOS ANGELES

Probably the best known resort area in northern Baja is Bahia de Los Angeles. Before completion of the paved highway, "L. A. Bay" was truly isolated and visited only by yachtsmen, pilots, and an occasional truck driver. Today's Highway 1 goes right to the area and new resorts have been built. The true pioneer of "L. A. Bay" is Antero (Papa) Diaz and his family who stayed on after the onyx and gold mines closed down. They erected a house, chapel, and a string of motel units. They provided charter fishing boats for visitors and generally made friends with everyone who landed on the sandy flight strip right in their front yard. (Fig. 12-12).

There was a small hump in the middle of the field that was the indirect cause of a freak, though fortunately a non-injury accident. A professional charter pilot has stopped to pick up a honeymooning couple. (This was back a number of years when L. A. Bay was an ideal honeymoon spot—no telephone, no bellhops, and the proprietor turned off the light plant at about 10:30 every night.) On takeoff, the honeymooners and their pilot rolled down the runway and up over the hump. At that go/no-go speed where the plane was traveling too fast to stop and not quite fast enough to fly, the pilot discovered that there had been a burro asleep on the flight strip. The engine noise woke him up and the burro trotted into the path of the plane. Result: one more washout, but no injuries to anyone but the burro.

A new 5000-foot paved runway, heading 33-15, has been built about a mile north of the existing 3000-foot dirt strip that goes right through the middle of the cluster of buildings that is "town" (Fig. 12-13). When the new airport is open, we were given to understand that the old one will be closed, thus depriving visitors of a most interesting spectator sport—watching first-time visitors trying to get their aircraft on the ground all in one piece. Such is the price of progress.

The Villa Vitta Resort has recently been opened at L. A. Bay by the operators of Jim's Air at Lindbergh Field, San Diego. In addition to a restaurant and "Charly Boy" bar, they have a large swimming pool, Jacuzzi, and sauna.

One of the isolated resorts far from the paved highway is Punta San Francisquito, some 55 miles south of Bahia de Los Angeles. It is accessible

Fig. 12-12. Ercoupe parks in the front yard at L. A. Bay. Rusting anchor in the foreground is used as an ornament for avid photographers.

Fig. 12-13. High altitude photo of L. A. Bay. Resort and old airport are at the left of the photo taken looking north. New airport nearing completion is in the center.

by four-wheeled vehicles on a narrow dirt road from El Arco, but this point is still "old Baja." A number of palm-thatched rooms have been built, the adjoining dry lake serves as an airport, and boats are available (Fig. 12-14). There's a store, and perhaps much more by now. It has been a number of years since we've landed here, but we'll try it again one of these days.

Check on the current status of accommodations at this resort before leaving the border by writing Punta San Francisquito, 2096 Newton Ave., San Diego, CA 92113. This same caution applies for any spot you plan to visit in Baja with the exception of the largest resorts. The motel may be closed, out of fuel, out of beer, or have its runway made unusable by rare tropical rains.

GUERRERO NEGRO THROUGH MULEGÉ

One of the more interesting stops in Baja is Guerrero Negro, location of the large salt mine. Actually, sea water is pumped into thousands of evaporating ponds. After the water has evaporated, huge trucks scoop up the pure salt and take it to a modern loading dock where it goes aboard barges and is carried to Cedros Island. There it is transferred to oceangoing freighters for delivery to Japan, Mexico, and the U.S. (Fig. 12-15).

Scammon's Lagoon just west of Guerrero Negro is the location of the annual migration of the California grey whale on a 6000-mile journey from the Bering Sea to bear their young. Whale-watching is at its best in January, February, and early March. Both Mexico and the U.S. have made the California grey whale a protected species. Seen from the air or the ground, the cavorting antics of these huge mammals are fun to watch. If you'll pardon the expression, they put on "a whale of a show."

You'll find at least two airports here, perhaps some fuel (but don't count on it), two small motels, and a 30-room El Presidente Hotel. There is a towering 135-foot high steel monument that marks the 28th parallel on the highway, signifying the separation between the State of Baja (northern part) and the Territory of Baja del Sur.

The small mission town of San Ignacio is worth visiting and has a fair dirt airport on a mesa just east of town. Last time we landed here, there were small rocks scattered around the runway and tiedown area. At least two other runways in various states of disrepair are nearby, so check ahead, then look carefully before deciding to land. If you buzz town, a taxi will show up at your landing spot. San Ignacio has many acres of green date palms, a welcome relief to the brown scrub bushes of the desert. The Jesuits

Fig. 12-14. Dry lake serves as a portion of the airport for the isolated resort at San Francisquito. Thatched-roofed buildings are erected near the shore.

Fig. 12-15. Large barges being filled with raw salt in the shallow lagoon at Guerrero Negro. This is reloaded into oceangoing freighters at nearby Cedros Island.

founded a mission here in 1728. The present church built by the Dominicans was completed in 1786. The lava rock walls are four feet thick and the building still serves as an active parish church. In the area on both sides of San Ignacio are many cave paintings by prehistoric Indians. Mule trips to these sites can be made at La Posada Motel. A new El Presidente Hotel has been opened 2-½ km west of the highway junction in town. San Ignacio is a

Fig. 12-16. Industrial copper mining town at Santa Rosalia. Large smokestack is built atop a hill to provide a good draft for the smelter.

Fig. 12-17. Northern end of the Santa Rosalia Airport with the town and smelter in the background. Because of its hillside location, some dangerous winds can exist.

historic town unto itself and differs completely from almost all the other towns in Baja that are on the shoreline.

Santa Rosalia is the next town along the gulf coast. It is a busy copper-mining town of 12,000 with narrow, congested streets (Fig. 12-16). Mining here dates back to a French company in the 1880s and the rows upon rows of uniform frame buildings look more like a company town than anything we've seen in Mexico. Even the church is constructed of flat galvanized iron. Historians note that this church, designed by A. G. Eiffel for the 1898 Paris Worlds Fair, was shipped around the horn to Santa Rosalia by mistake.

The airport just south of town is on a small hill with the runway cut out of the side of the rock (Fig. 12-17). In high wind conditions, it can be extremely challenging and most visiting pilots continue another 35 miles to the tropical oasis of Mulegé.

On the Pacific Coast side abeam Santa Rosalia is the small fishing village of Punto Abreojos (about 500 people) with a store, a tiny cafe, gasoline drums, and commercial radio communications. The airport just north of town is used by DC-3s to carry abalone and lobster to market.

Mulegé is considered to be one of the true garden spots in Mexico. Before completion of the paved road, it was a very sleepy village along the banks of the Rio Mulegé (Fig. 12-18). Date farms vied with figs, oranges, bananas and olives as local produce. Most visitors came in by air. That has now changed as hordes of campers and trailers have invaded the area. We've waited at the gas pit at Hotel Serenidad while lines of motorcyclists topped off with av/gas. Other hotels include Las Casitas and Old Hacienda Mulegé in town and Hotel Mulegé on a bluff east of town on the north side of the river (Fig. 12-19). Hotel Mulegé has two short flight strips adjoining, but as this book is being prepared, most air traffic lands at Serenidad.

The town itself has extensive facilities for this far out in the boonies, including a supermarket, laundromat, gift shop, telephone, and telegraph. Streets are dirt, narrow, and one-way. The Territorial Prison overlooks the town on the north and is open to visitors during the day when prisoners, if any, are permitted to work in town. The mission founded in 1705 and completed in 1766, has been restored by the government and is a popular picture-taking location.

Mulegé, of course, has great fishing (Fig. 12-20). The broad Bahia Concepcion just south of Mulegé offers sheltered waters for the many sport fishing boats that can be chartered. Mulegé is as tropical as they come, but has so far remained a simple town. As of 1981, there wasn't a high rise building in sight. A day and night spent here gives the fly-in tourist more of the flavor of earlier, unsophisticated days in Baja than just about any other spot along the peninsula.

Fig. 12-18. Airports at the mouth of the Mulegé River. Most used airport today is at Hotel Serenidad on the far south side of the river. Flight strip in the foreground for Hotel Mulegé is not now in use.

Fig. 12-19. Flight strip located behind the Hotel Mulegé. Note palm trees and desert oasis along the river in the foreground.

LORETO AND LA PAZ

Loreto, next of the major towns on the Gulf Coast, is only a short 75 mile flight. The oldest mission in either Baja or "Alta" California, dates back to 1697. For the next 132 years, Loreto was the capital of Baja

Fig. 12-20. Aero Commander twin lands (downwind) in front of the Hotel Serenidad at Mulegé. White rocks lining the boundary of the airport proved to be more of a problem than they were worth and were removed.

Fig. 12-21. New Hotel El Presidente just south of Loreto as seen from the air. This hotel is being advertised all over Mexico.

California. Loreto has long been a destination for pilots. Ed Tabor's Flying Sportsmen Lodge once had a flight strip adjoining, but the advent of the new Government airport south of town put this landing area in jeopardy. In addition to the Flying Sportsmen, Loreto had the quiet, friendly 33-unit Hotel Oasis a half-mile south of the town plaza. Latest and largest is the El Presidente Hotel at Playa Nopolo south of town and south of the airport (Figs. 12-21, 12-22). This new resort complex was recently developed by

Fig. 12-22. New Presidente Hotel has two swimming pools, an underground disco and just about anything that a resort hotel can offer.

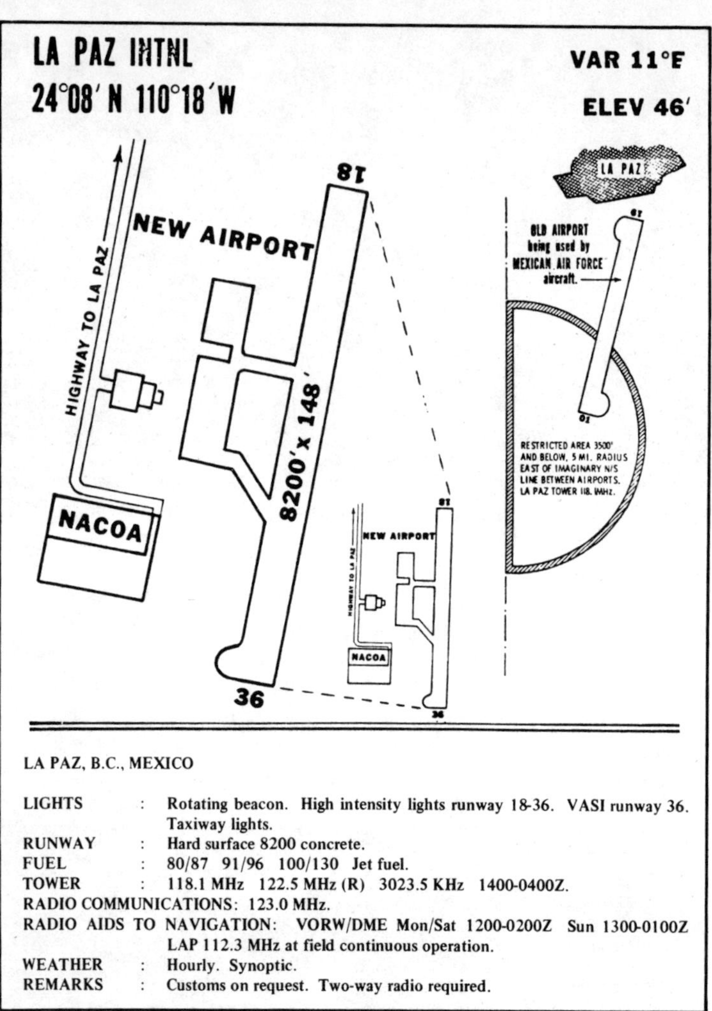

LA PAZ, B.C., MEXICO

LIGHTS	:	Rotating beacon. High intensity lights runway 18-36. VASI runway 36. Taxiway lights.
RUNWAY	:	Hard surface 8200 concrete.
FUEL	:	80/87 91/96 100/130 Jet fuel.
TOWER	:	118.1 MHz 122.5 MHz (R) 3023.5 KHz 1400-0400Z.

RADIO COMMUNICATIONS: 123.0 MHz.

RADIO AIDS TO NAVIGATION: VORW/DME Mon/Sat 1200-0200Z Sun 1300-0100Z LAP 112.3 MHz at field continuous operation.

WEATHER	:	Hourly. Synoptic.
REMARKS	:	Customs on request. Two-way radio required.

Fig. 12-23. Drawing of the airport layout at La Paz. Note mention of the military airport now used by the Mexican Air Force.

the Mexican Government and has been the subject of extensive promotion throughout Mexico. When we had lunch recently at the Mexico City Airport, the paper tablemats were telling about this new Loreto Hotel.

There's a great deal of very little along the coastline between Loreto and La Paz once the highway turns inland. Broad beaches give way to rugged cliffs that run right down to the water. Some single-engine pilots on their first trip may want to stay with the highway and fly perhaps 50 miles farther, but that's a personal option. Don't be confused and use the military airport that is closer to town than the huge jetport (Fig. 12-23). La Paz is a comfortable large town of over 60,000 and is the governmental capital of Baja California Sur. The town faces northwest toward the bay and produces some of the most inspiring sunsets to be found anywhere. La Paz is a free port with many imported items available. The palm-lined *malecon* (sea wall) makes for a fine sunset photograph.

There is a wide range of hotels from the new El Presidente two miles southwest of town to several smaller downtown hotels. Largest of the downtown hotels is Los Arcos, a sprawling series of buildings with 193 units, including 16 garden bungalows and a three-story wing (Fig. 12-24).

La Paz can be a complete destination by itself or a jumping off spot for "the tip." No attempt is made here to visit all the fly-in resorts between La Paz and Lands End at Cabo San Lucas. We will, hwever, mention several that we have visited personally over the years. Hotel Punta Arenas has 40 deluxe suites, all with ocean views. There's a 4200-foot runway 29 D.M.E. on the 91° radial from the La Paz VOR, Unicom is 122.8.

AROUND THE TIP OF BAJA

Farther down the peninsula is Punta Pescadero, a delightful hideaway spot sheltered from the paved road by 10½ miles of rugged, unpaved approach road. Unless you're a dedicated driver, you reach Punta Pescadero in your own plane or a chartered air taxi. The 2900-foot flight strip has clear approaches and good tiedowns. As with most other top resorts toward the tip of Baja, reservations are recommended. We spent one Christmas Holiday here on a mattress rolled out on the pool table in one of the maintenance buildings, but it was worth the trouble. General Managers are Cathie and Bill Ward and stateside reservations and information are handled at P.O. Box 103, Santa Maria, California 93454 (Figs. 12-25, 12-26).

Fig. 12-24. Waterfront in downtown La Paz. Several good commercial hotels can be found in this area.

Fig. 12-25. Between Punta Pescadero and Rancho Buena Vista is Bahia de Palmas with its 2000-foot landing field parallel to the beach.

Just a few miles farther down the coast is Rancho Buena Vista where the Walters family have been hosting deep sea fishermen for two generations. During the filming of *Tigers of the Sea,* we flew a Piper Aztec in and out of the Rancho Buena Vista dirt strip on several occasions shuttling passengers in and out from Mexico City (Fig. 12-27). That overwater flight direct from Rancho Buena Vista to Guadalajara is fine with a good twin, but not recommended for a single-engine pilot. You're out of sight of land for many, many minutes at midpoint, with search and rescue just not available.

Fig. 12-26. Here's what the Bahia de Palmas flight strip looks like on final approach. Field is 2000 feet long. Don't undershoot and land in the sand.

Fig. 12-27. Piper Aztec used by the author in filming *Tigers of the Sea* is parked at the Rancho Buena Vista flight strip.

We asked veteran pilot Charles M. Walters of Rancho Buena Vista to share some thoughts on how flying in Baja California has improved. Here's what he had to say:

"Over the years I really don't see too much change in private flying here in Baja with the exception of perhaps more of it. Our neighbor, Las Palmas, has gone all out for pilots and they have more and more groups. I find that for our business, we must go after the fisherman, for he spends more time with us. The fly-in groups as a rule spend about three to four nights and are not usually serious fishermen. We have about 60% of our business geared to fishing and thus must go after it. Pilots today are a younger and more affluent group than in the past. For example, three Cessna 411s left just yesterday; the average age of the pilots was about 32. They fished one day and spent three with us. Fun in the pool and sauna tub the first day; quieter the next two. Orange County people in the building trade.

"From my own viewpoint, I feel that Mexico has greatly changed its rather hostile and suspicious attitude toward the pilot and today it is a very comfortable but slow experience covering the check in and gassing at borders. There is not the hand-out attitude, but for me I still believe a few dollars means better treatment and worth it. The recent change of having to land at the nearest International Airport has made it slower checking through.

"We have had but one bit of vandalism since we have been in business, and this was since the highway went in. Other than this, the highway has made it easier to get damaged planes out, parts in, and no doubt promote more flying due to the poor condition that road has been in for the past three years or so. It is murder to want to drive it. The huge tracks and Greyhound-type buses roar down the peninsula making the holes bigger and bigger!"

You'll usually find Walters' Beech Baron 55 parked near the Rancho Buena Vista flight strip.

Another of the many fine resorts approaching the tip is Punta Colorada, complete with its own 3300-foot airstrip a half-mile drive from the 26-unit

Fig. 12-28. Town of San Jose del Cabo as seen from the air. New jetport is some five miles to the north.

hotel. This resort is also isolated 13 miles from Highway 1 as is the smaller seven-unit Laguna Guesthouse on Punta Arena Bay just south of Punta Colorada.

If you're in the mood for sightseeing, two inland towns of Santiago and Miraflores have dirt flight strips. Look them over carefully from the air for rocks, bottles, ditches, and cattle before attempting a landing. Pilots of tri-geared aircraft should remember that unpaved flight strips can cause propeller dings.

The new jet-sized San Jose del Cabo Airport is seven miles inland and is geared to handle the large numbers of passengers headed for the tip. We've flown over it and admired the long runways, but really had no reason to land and then take a cab or bus to the resorts on the tip. Each of the larger resorts either has its own flight strip or uses a small one that's much closer than the international field (Fig. 12-28). The town of San Jose del Cabo has about 10,000 people and is the largest town south of La Paz. There are a number of stores and a small hospital. At one time there was a short, sandy flight strip adjoining the town cemetery, but it was poorly maintained and marginal. Probably it has long since gone (Fig. 12-29).

Hotel Palmilla is one of the oldest, yet one of the nicest of Baja's luxury resorts. It is just 1½ miles west of the town of San Jose del Cabo and has a 5000-foot lighted airstrip (Unicom 122.8) (Fig. 12-30). Since the airport has quite a slope toward the sea, most landings are made uphill and takeoffs down over the cliffs. Palmilla has all the graciousness of a fine hotel with a

Fig. 12-29. New Presidente Hotel is shown under construction on the beach at San Jose del Cabo.

putting green, driving range, and tennis court. For those who aren't fishing, there is hunting and horseback riding. Fishing cruisers with skipper and boatman are available as are smaller skiffs.

Adjoining Palmilla is the Hotel Cabo San Lucas, also with a 3600-foot flight strip and Unicom on 122.8 (land uphill) (Fig. 12-31). With both airports on the same frequency, there's less chance of an accident. For many years, the Hotel Cabo San Lucas has been the plushiest resort on the tip. It is expensive, but well worth it. Rooms are designed with colonial

Fig. 12-30. Cessna 310 flies over the Palmilla resort. Note famed bell tower of the resort and well-maintained flight strip at the right. Airport slopes uphill away from the water.

Fig. 12-31. Famed Cabo San Lucas Hotel with adjoining airport. This resort is considered by some to be the finest hotel in Mexico. Compare this photo with 12-2 to see the changes in 20 years (courtesy Mexican Tourist Council).

styling and furniture built right on the property. There is even a private yacht anchorage. A staff of over 280 bilingual employees takes care of the 125 guest rooms. The year-round climate, relatively free of humidity, is 80-85°F in the day and 70-72°F at night. There's a glass-bottom boat available as well as cars and dune cycles. Ray Cannon, author of the bestselling *Sea of Cortex*, said "This is the most elegant resort hotel in all of Mexico, if not the world . . . its beauty is surpassed only by its service and cuisine."

Fig. 12-32. Small town of Cabo San Lucas with its deepwater harbor as seen from the air. Hotel Finisterra is at the bottom left (courtesy Mexican Tourist Council).

Fig. 12-33. Ferry from Puerta Vallarta docks at Cabo San Lucas. Photo taker from the Hacienda Hotel.

It is difficult-to-impossible to pick what might be the best hotel on the tip since new units are blossoming almost every month. Thus, any report in 1982 can well be out-of-date next year.

The small town of Cabo San Lucas (Fig. 12-32) itself has a small deepwater harbor and is the terminal for the ferry that takes 22 hours to Puerto Vallarta (Fig. 12-33). The bay was a frequent hiding place for pirates after Spanish treasure in the 16th and 17th centuries. A small but aromatic fishing cannery shares the harbor. Recently, a government-sponsored regional arts center makes and sells distinctive black coral jewelry.

Fig. 12-34. Hotel Hacienda at Cabo San Lucas undergoing expansion for another wing of rooms (courtesy Mexican Tourist Council).

Fig. 12-35. Famed rock formations at the very tip of Baja California as seen from a glass-bottomed boat. Beyond the arch is the Pacific Ocean.

As of this writing, hotels in Cabo San Lucas include the striking Finisterra. We were with Gunnell's "Fly for Funsters" aviation group when they officially opened that hotel a number of years ago.

The Hacienda Hotel overlooks the Gulf de California (Figs. 12-33, 12-34) while the economy 30-unit motel Mar De Cortez is near the center of town. You can choose the Hotel Solmar on the beach at the southernmost tip of the peninsula. Both the Hyatt Cabo San Lucas and Twin Dolphin Hotel are east of the town center and within a half mile of each other. These latter two were under construction as this report was completed.

The allure of the tip of Baja is so strong that things are changing rapidly—*very* rapidly for Mexico. Each new visit will show new resorts, new flight strips, and new tourist facilities (Fig. 12-35). The trip to the tip of Baja represents a long flight in an economy light plane—776 air miles. The required border-crossing stops and the new flight plan fees, a substantial raise in ASA fuel prices and the overall inflation in food and lodging cause a considerable investment in both time and money. As an example of fuel costs, our latest 1981 trip to the tip of Baja cost an even $300 for 100-octane av/gas (19 hours), the same amount of money we spent just one year earlier in flying the same airplane all the way to Acapulco and back (29 hours).

Baja is a peninsula all unto its own. It has undoubtedly the best fishing in the world. It has colorful, secluded bays with some of the whitest sand you'll ever see and crunch your bare feet in. It has a colorful history dating back to pirates and missionaries. Today it has some of the largest jet-set resort hotels to be found anywhere. When you fly yourself to Baja, you'll find it a rewarding experience!

Chapter 13

U.S. Border Crossing

Crossing the border back into the United States should be no problem. After all, you're coming back home. To keep your home-coming as pleasant as you'd hoped, stick with the rules of the U.S. Customs.

Proof of citizenship is far more strict for U.S. citizens returning to their homeland than it is for them to enter Mexico. We strongly recommend that you carry a U.S. passport, even an outdated one, while traveling anywhere outside the U.S., even though this identification is not required in Mexico. If you don't have a passport, a certified copy of your birth certificate will do. Check with Customs officials for other alternatives prior to departure from the U.S.

ADVANCE NOTICE REQUIRED

Be sure to give U.S. Customs at least 15 minutes prior notice before landing at a U.S. International Airport of Entry. Your "ADCUS" (advise Customs) flight plan should be transmitted in the air before your final landing at a Mexican airport of entry/departure to turn in your last flight plan and your ever-loving, usually well-worn Mexican Declaration, Form GHC-001. Depending on the price of fuel and the amount of space in your tanks, you may also want to top off before departing from Mexico.

At most places along the Mexican Border, these flights inbound to the U.S. are very short. Tijuana and Brown Field, for example, are within each other's traffic pattern. Mexicali and Calexico are less than 20 miles apart, as are Brownsville and Matamoros. If you plan to be on the ground at your last Mexican stop for any length of time, you can either telephone personally or make your in-flight filing with enough delay to take care of your ground time.

We urge that you take care of this flight plan in person as there is a $500 fine for failure to notify. On one occasion at Tijuana, we were assured that the Mexican officials would take care of advising Customs and notifying Brown Field of our flight plan. We just happened to learn before leaving that the phone lines had been down all day between the Mexican Border and the U.S., and it would have been impossible for the contact to be made. We went out to our plane and notified them ourselves by radio.

The *U.S. Customs Guide for Private Flyers* (General Aviation Pilots) 1980 edition spells out the reasons for providing advance notification this way:

"In order to have an officer present to provide Customs service for you and your aircraft, Customs must be notified of your intention to land and time of arrival or penetration. The notification may be provided through FAA however, this entails the relaying of information and is not as timely or reliable as direct communication. It is recommended that, if possible, pilots attempt to communicate directly with Customs by telephone or other means to insure that an officer will be available at the time requested. It is the ultimate responsibility of the pilot to insure Customs is properly notified, and the failure to do so may subject the pilot to penalty action."

Special reporting requirements for the southern border are detailed thusly: "All pilots of private aircraft arriving from a foreign place in the western hemisphere south of 33 degrees north latitude which cross into the U.S. over a point on the U.S. border between 95 and 120 degrees west longitude are required to communicate to Customs by telephone, radio, or other means either directly or through the FAA Flight Station, their intention of landing and the intended point and time of border crossing not less than 15 minutes prior to crossing the border.

"Due to unreliable communications relay from Mexico, a flight plan filed in Mexico may not be transmitted in time to meet this reporting requirement even though the flight plan included 'ADCUS' or 'Advise Customs.' Pilots are advised to rely solely upon telephone or radio communication to notify Customs of intended arrival on a timely basis.

"The report to Customs must include the following:

- ☐ Type of aircraft and registration number.
- ☐ Name of aircraft commander.
- ☐ Number of United States citizen passengers.
- ☐ Number of alien passengers.
- ☐ Place of last foreign departure.
- ☐ Other countries visited.
- ☐ Estimated time and location of crossing U.S. border.
- ☐ Name of nearest intended U.S. airport of first landing.
- ☐ Estimated time of arrival.

"All private aircraft, unless exempted, are required to land at one of the specially designated airports listed below."

We have included the phone numbers for your convenience. More detailed information regarding requirements of the various U.S. Government agencies involved is available in the International Flight Information Manual, normally available at FAA Flight Service Stations. It is recommended that all aircraft entering the United States contact the associated Flight Service Station 15 minutes prior to crossing the border to assure Customs notification.

DESTINATION AIRPORT OF ENTRY	*24-HOUR CUSTOMS PORT*	*ASSOCIATED FSS*	*CUSTOMS PHONE NUMBER*
Brownsville International	Brownsville, TX	McAllen	512-542-4232
Calexico International	Calexico, CA	Imperial	714-357-1195
Del Rio International	Del Rio, TX	San Antonio	512-775-8502
Bisbee-Douglas International	Douglas, AZ	Douglas	602-364-8486
Eagle Pass Airport	Eagle Pass, TX	San Antonio	512-773-9468
El Paso International	El Paso, TX	El Paso	915-543-7430
Laredo International	Laredo, TX	San Antonio	512-722-9551
Miller International	Hidalgo, TX	McAllen	512-843-2231
Nogales International	Nogales, AZ	Tucson	602-287-2562
Presidio - Lely (daylight hours only)	Presidio, TX	El Paso	
Brown Field, San Diego, CA	San Ysidro, CA	San Diego	714-428-7206
San Diego International (Lindbergh Field)	San Ysidro, CA	San Diego	
Tucson International	Nogales, AZ	Tucson	602-287-2562 602-792-6359
Yuma International	San Luis, AZ	Yuma	602-726-2601 (FAA Flight Service) 602-726-2595 (San Luis)

Pilots wishing to use airports other than the 14 listed above should check with U.S. Customs about other "international" or landing rights airports. Such airports usually require 30 days prior notification in writing and are specialized in requirements.

If an emergency landing is made in the United States, the pilot should report as promptly as possible by telephone or most convenient means to the nearest Customs office. He should keep all merchandise or baggage in a segregated place and should not permit any passenger or crewmember to depart the place of arrival or comingle with the public without official permission, unless it is necessary for preservation of life, health, or property.

PENALTIES FOR VIOLATIONS

U.S. Customs law provides for "substantial penalties for violations of the Customs regulations and aircraft operators and pilots should make every effort to comply with them. Examples of the more common violations and resulting penalties include:

- ☐ Failure to report arrival—$500.
- ☐ Failure to obtain landing rights—$500.
- ☐ Failure to provide advance notice of arrival—$500.
- ☐ Failure to provide penetration report on southwest border —$500.
- ☐ Departing without permission or discharging passengers or cargo without permission—$500.
- ☐ Importation of contraband, including agriculture materials or undeclared merchandise, can result in penalty action and seizure of aircraft which varies according to the nature of the violation and pertinent provision of law.

"If a penalty is incurred, application may be made to the Customs officer in charge for a reduction in amount or cancellation, giving the grounds upon which relief is believed to be justified. If the operator or pilot desires to further petition for relief of the penalty, he may appeal to the appropriate District Director of Customs. If still further review of the penalty is desired, written appeal may be made to the proper Regional Commissioner of Customs and, in some cases, to Customs Headquarters."

During the research for this book, we made several border crossings through Brown Field, just across the border from Tijuana, B.C. At this time, there were so many $500 fines being levied against inbound U.S. aircraft for failure to provide advance notice of arrival that the Customs office in San Ysidro sent one of the Inspectors, also a private pilot, named Bert Smith to the airport office on a 30-day assignment to see what could be done to cut down on these inadvertent violations.

We met U.S. Customs Inspector Bert Smith on what was to have been a routine border crossing. Well, it was *almost* routine. Tijuana had a temporary telephone failure so their Operations Desk could not dispatch our outbound flight plan from Mexico. We solved that by talking with Brown Field Tower from the Tijuana airport parking lot. We asked the tower to relay our flight plan and notify Customs. Then we waited 30 minutes before taking off for Brown Field.

We waited beside the airplane as the big billboard advises until the Customs man appeared, talked with us briefly, and had us follow him into the office. We showed him our passports as we always do. Inspector Smith picked up our passports and disappeared into the back room for three or four minutes. When he returned, he asked us for the "N" number of our Cessna. We replied "N2672D," and pointed out the window to the small numbers on the tail. The Inspector nodded, left the room again and when he returned, things were a good bit more cordial. It seems that somewhere in transmission between Brown Tower, the FAA, and Customs our call numbers had been transposed. When the Inspector had put the transmitted numbers into the computer, it said "tilt," or some such rejection.

Inspector Smith then explained his temporary assignment to smooth up the problems at Brown Field. As a result of his efforts, better communications have been established between the Operations Desk at Tijuana with a line directly to the U.S. Customs office in San Ysidro. A permanent assignment of a Customs man ten hours a day has been instituted at Brown Field and inbound pilots are finding it simpler and quicker to clear Customs at the low-activity Brown Field rather than going through the Terminal Control Area (TCA) to reach San Diego's Lindbergh Field.

Now U.S. Customs officials assigned to Brown Field are taking visits to Tijuana operations, frequently by lightplane, so that they can better understand the problems on both sides of the border. Many of the $500 violations are now changed to warnings.

While it is no panacea for inbound flight notification problems, it is always a good idea to keep a copy of your inbound Mexican flight plan with the Comandante's signature and stamp. This will serve as proof that you did, in fact, file a flight plan. However, it is still the responsibility of the pilot to make sure that U.S. Customs is notified. If there is any doubt on your initial south-of-the-border contact with the FAA/FSS on your "ADCUS" border crossing flight plan take the time and the 20¢ or whatever to call U.S. Customs before leaving Mexico.

PURCHASES

Customs men and women are well trained and know a great deal about the habits of returning U.S. citizens. The type of shoes you wear will give a good indication of where you live. Your list of purchases gives additional insight. If, for example, you bring in $300 worth of cotton shirts (normally a high-duty item), there will be little question if they're the proper size for you or a reasonable assortment for gifts. If the Inspector is suspicious, however, he may have someone in your hometown area check to see if you're suddenly in the business of selling shirts.

We landed at Brown Field one sunny afternoon, taxied up to the Customs gate, and shut down. The Inspector who greeted us seemed preoccupied. He took a quick look inside the Cessna, asked a cursory question or two in the office and ushered us out the door with a brief comment, "I can fill the remaining blanks in later."

As we walked back to our ship, a low-winged six placer was just shutting down. Promptly two Inspectors and their not-so-friendly dog met the aircraft. After the passengers disembarked, all the luggage was taken into the office. The dog was "sniffing" and one inspector was laboriously searching the aircraft as we started up and taxied out.

If you are ever the slightest bit tempted to "get by" with undeclared items, whether it be jewelry or contraband, remember that the person who sold you that high-priced item in Mexico stands to gain half of any fine you may incur by attempted smuggling, so chances are good that your expensive purchase, no matter where it was made, has been reported and that information is perhaps waiting for your arrival at the border. In the case of narcotics, we have heard, (but not substantiated) that the snitch not only gets 50% of your fine, but also gets his dope back as well!

If you are a "good guy" and seem to draw more than a normal interest from Customs Inspectors, you might check into the pedigree of your airplane. If you've recently purchased it, there is always a chance that it is still listed on the Government's list of "questionable" aircraft and that N number will print out every time it is entered into the computer.

When AOPA Vice President Bob Warner and his partner Jeff Gilley, AOPA Director Airport Operations, purchased a used Cessna 182 a number of years ago, their first trip or two through U.S. Customs were really rough. Finally the new owners, pilots knowledgeable with the ways of the aircraft world, asked a Customs man at the border why they were continually getting such a hassle. The reply was that their aircraft was on the list of previous violators. Once the change in ownership had been established, the new owners had no more unusual scrutiny.

REPAIRS TO PRIVATE AIRCRAFT

Aircraft belonging to a resident of the U.S. and taken abroad for noncommercial purposes and returned by the resident shall be admitted free of duty upon being satisfactorily identified. Repairs made abroad to such aircraft, if incidental to use abroad, must be reported to Customs but are not subject to duty. Repairs not incidental to use abroad and alterations and additions made abroad shall be assessed with duty up to their value at the rate at which the aircraft itself would be dutiable if imported. Accessories acquired abroad are dutiable as if separately imported.

In the event that foreign-made parts, such as spare engines, must be replaced abroad, these parts may be dutiable upon reentry to the United States. The reimportation of U.S.-made parts may require entry. It is advisable that the nearest U.S. Customs office be contacted for clarification of this matter in each particular instance.

DO'S AND DON'TS

A Customs pamphlet titled *Know Before You Go* (Department of the Treasury, U.S. Customs Service, Washington, D.C. 20229) (Fig. 13-5) carries the latest rules and regulations on exemptions to duty, gifts, penalties and specific articles that are considered either free or dutiable. Since these regulations change from time to time, the authors strongly urge that you obtain the latest available copy of this small pamphlet before departing from the U.S.

The following is a guideline list of "Do's and Don't's" for quick reference written by the United States Custom Service:

- ☐ Do read Customs hints. Know your Customs rules.
- ☐ Do get Customs information from Customs officials.
- ☐ Do declare every article acquired abroad and accompanying you, including gifts.
- ☐ Do keep a record of all acquired items, save sales invoices, and convert prices to U.S. currency to save time.
- ☐ Do pack your purchases and gifts in one piece of luggage.
- ☐ Do have all your baggage ready for inspection.
- ☐ Do know your State as well as Federal liquor restrictions.
- ☐ Do understand every country insists upon a thorough Customs examination for returning residents.
- ☐ Do realize we want to help you clear Customs quickly.
- ☐ Do understand we appreciate your patience. Remember Customs inspectors have a difficult task to fulfill.
- ☐ Don't exceed your Customs exemption without expecting to pay duty.
- ☐ Don't forget your purchases sent home are subject to duty.
- ☐ Don't be surprised if we open all your luggage.
- ☐ Don't forget all accompanying foreign purchases must be declared, even those you have worn or used.
- ☐ Don't accept the offer of a "false" sales invoice. This could result in seizures and penalties.
- ☐ Don't rely on the "experienced" traveler or foreign seller for your Customs information.
- ☐ Don't bring back lottery tickets.
- ☐ Don't bring back fruits, plants, vegetables or meat without permits or advice from the Department of Agriculture.

OVERTIME CHARGES

At the time this book is completed, U.S. Customs overtime charges apply up to a maximum of $25 for services provided Sundays and holidays before 8 a.m. and after 5 p.m. On a recent trip back from Mexicali, we cleared U.S. Customs at Calexico, California, shortly before 5 p.m. on a Sunday. As we waited and talked with the Inspector, it became 5 p.m. and a couple of stragglers arrived. As the Inspector filled out his paperwork, he asked the pilots for their landing time. When both pilots said "five minutes after five," the Inspector reminded them of the overtime charges.

Inbound pilots who have filed for landing at two nearby U.S. International Airports because of weather or other factors are cautioned to cancel the request for overtime service at their original destination when diverting to a different airport so they won't have to pay two sets of overtime charges. Such a situation could occur in San Diego, for example, if there were a question of being able to get into either Brown or Lindbergh because of weather.

There is considerable pressure being applied at the Washington D.C. level by AOPA and other aviation groups to establish an increased number of airports that can provide free 24-hour Customs inspection for general aviation aircraft. AOPA believes that if 24-hour per day Customs inspection services were available at selected general aviation airports, the demand for overtime inspectors of general aviation aircraft at the major hub airports would decrease considerably. Concurrently, AOPA is battling a proposal to increase the overtime inspection fee to $50 and later to whatever fee could be justified by Customs Service.

DEPARTMENT OF THE TREASURY
UNITED STATES CUSTOMS SERVICE

PRIVATE AIRCRAFT INSPECTION REPORT
6.2, 6.14, C.R.

PARTIAL LIST OF COUNTRY CODES

AC	ANTIGUA	GJ	GRENADA	PQ	CANAL ZONE
BB	BARBADOS	GP	GUADELOUPE	SC	ST. CHRISTOPHER-NEVIS-ANGUILLA
BD	BERMUDA	GT	GUATEMALA	SQ	SWAN IS.
BF	BAHAMAS	GY	GUYANA	ST	ST. LUCIA
BH	BRITISH HONDURAS	HA	HAITI	TD	TRINIDAD & TOBAGO
CA	CANADA	HO	HONDURAS	TK	TURKS & CAICOS IS.
CJ	CAYMAN IS.	JM	JAMAICA	VC	ST. VINCENT
CK	COCOS IS.	MB	MARTINIQUE	VE	VENEZUELA
CO	COLOMBIA	MX	MEXICO	VI	BRITISH V.I.
CS	COSTA RICA	NA	NETHERLANDS ANTILLES	VQ	V.I. (U.S.)
DR	DOMINICAN REPUBLIC	NU	NICARACUA		
EL	EL SALVADOR	PN	PANAMA		

M — 1. SYSID NR. — 2 0 :

M N — 2. A/C TAIL NR. — 3 1 :

M N — 3 8 — 3a. PILOT LAST NAME — 3b. PILOT FIRST NAME

M N — 6 6 :

4. LAST FOREIGN AIRPORT OF DEPARTURE— (a) city — (b) country code

38 Cont.

5. AIRPORT CODE — 6. ARRIVAL TIME — 7. TODAY'S DATE (MMDDYY)

8. PILOT D.O.B. — M M D D Y Y — 9. PILOT LICENSE NR.

66 Cont.

N — 10. A/C MAKE — 6 1 :

61 Cont. — A C F T — 11. A/C COLOR(S)

PAIPS ENTRY (Customs Use Only)
a. Station Code
b. Date
c. Time

12. INSPECTOR SIGNATURE

PASSENGERS

LAST NAME	FIRST NAME	D.O.B.

Customs Form 178 (7-27-76)

PRIVACY ACT NOTICE
This information is provided pursuant to Public Law 93-579 (Privacy Act of 1974).

The information is requested under the authority of 5 U.S.C. 301; Reorganization Plan 1 of 1950; Treasury Department Order No. 165, revised, as amended.

Providing the information is voluntary; however, failure to provide the information requested may result in a delay in your aircraft being inspected.

The purpose of the information collected is to document the arrivals of general aviation type aircraft from foreign countries in order to administer Customs and related laws.

The routine uses which will be made of the information are as follows: (1) disclosure of the information to law enforcement agencies, (2) used as an intelligence gathering source, (3) Customs management functions such as measuring workload of Customs officers performing aircraft arrival inspections, and (4) disclosure to officers and employees of Customs and Treasury who have a need for the records in the performance of their duties.

GPO 964-410 ☆U.S. GOVERNMENT PRINTING OFFICE: 1978-717-595/503

Customs Form 178 (7-27-76)

Fig. 13-1. Sample U.S. Customs form #178 that is filled out by the Inspector on duty for each inbound U.S. aircraft.

PAPERWORK AT THE BORDER

When you and your passengers go into the U.S. Customs office at an International Airport, the Inspector has two pieces of paper to fill out. One is Customs Form 178 (Fig. 13-1), a Private Aircraft Inspection Report that lists the pilot's name, date of birth, last foreign airport, pilot's license number, aircraft make, tail number, color, plus a list of all passengers and their date of birth.

The second form required is I-92A (Figs. 13-2, 13-3), a report of private aircraft arrival. This form lists both the name, address and nationality of the pilot and also the aircraft owner. Departure from foreign soil and landing on U.S. territory are recorded as well as the names and immigration status of all passengers. The number of pieces of baggage is also required on this form. These forms can be requested in advance and filled out prior to landing in the U.S.—a recommended procedure that can save a little time at the Border.

The quickest way for the Inspector to fill out these forms is to ask the pilot for his license and copy the information from that form. The inspector may also ask for the pilot's medical certificate and the aircraft registration certificate.

Crew and passenger baggage will be examined in the same manner as that of other international travelers. A verbal declaration of articles acquired abroad will suffice, except that a written declaration, Customs Form 6059-B (or appropriate substitute), shall be presented when duty is to be collected or when the inspecting officer deems a written declaration necessary. Noncommercial cargo and unaccompanied baggage carried on board private aircraft shall be accounted for on a baggage declaration (CF 6059-B) prepared by the pilot in command, and appropriate entry for same shall be

U.S. Department of Justice • Imm & Natz Service

(PLEASE PRINT AND FILL IN ALL SPACES)

AIRCRAFT NO.:

MAKE: COLOR:

COMPLETE THIS PART ONLY IF U.S. BASED AIRCRAFT

Departed from U.S. on: ______ (Date)

at: ______ (U.S. Port/ City, State)

FOREIGN ITINERARY (Aircraft)

Departed for U.S. at: ______ (Hour)

From: ______ (Last Foreign Port / City, State)

Aircraft Arrived in U.S. at: ______ (U.S. Port / City, State)

on: ______ (Date) at: ______ (Hour)

Pilot — Name, Address & Nationality

Owner — Name, Address & Nationality

ESTIMATED TIME AND DATE OF ARRIVAL	D.	
	T.	
INSPECTION DATA	HOURS	MIN.
TRAVEL TIME		
WAITING TIME		
INSPECTION TIME		

FORM I-92A (REV.) 3-10-77 Y REPORT OF PRIVATE AIRCRAFT ARRIVAL

Fig. 13-2. Front side of the inbound aircraft form I-92A. These forms may be obtained in advance and filled out prior to arriving at the U.S. Customs station.

(REMOVE CARBON PAPER BEFORE FILLING IN BELOW)			
LIST OF PASSENGERS NAMES		IMMIGRATION STATUS	TOTAL INSPECTED
BAGGAGE — NO. OF PIECES		SMALLPOX VACCINATIONS PILOT ☐ PASSENGERS NO.____	
CARGO YES NO	INTERCEPTIONS YES NO	Signature of Inspecting Officer	

Fig. 13-3. Reverse side of Form I-92A lists passengers' names and number of pieces of baggage. Depending on a variety of circumstances, you may be required to open your baggage compartment and any of the luggage on board.

required. Customs officers will furnish the necessary forms. In addition, the inspecting officer may require that all personal items, baggage, and cargo be removed from the aircraft for inspection, and he may physically inspect the aircraft. It is the responsibility of the pilot to assist in opening baggage and compartments.

If your oral declaration is "in line" and your manner doesn't suggest that you're anything but honest, the complete inspection procedure, including required paperwork, should take no more than 10 to 15 minutes for four people.

Then you can heave a small sigh of relief to be clear of the red tape, climb back into your airplane, and head for home.

BUEN VIAJE!

We'd like to sign off this book the same way we did the *AERO* magazine article, but we were told that wasn't quite right. Somewhere north of Puerto Vallarta, we were in the cockpit of an Aeronaves 727 and the Captain was reading our low-altitude report in *Aero* with interest. Below us by 33,000 feet were the same beaches that we'd toured in the Cessna 170B. The article concluded:

"From the treatment we received on this, our latest trip to Mexico, we can say you can take your aircraft south of the border with few if any problems. Bien Viaje."

The jet Captain handed the magazine back and said, "You've made just one mistake in the whole story. It's not 'Bien Viaje;' it's 'Buen Viaje!'"

Appendix A: More Mexican Airports

Appendix A: More Mexican Airports

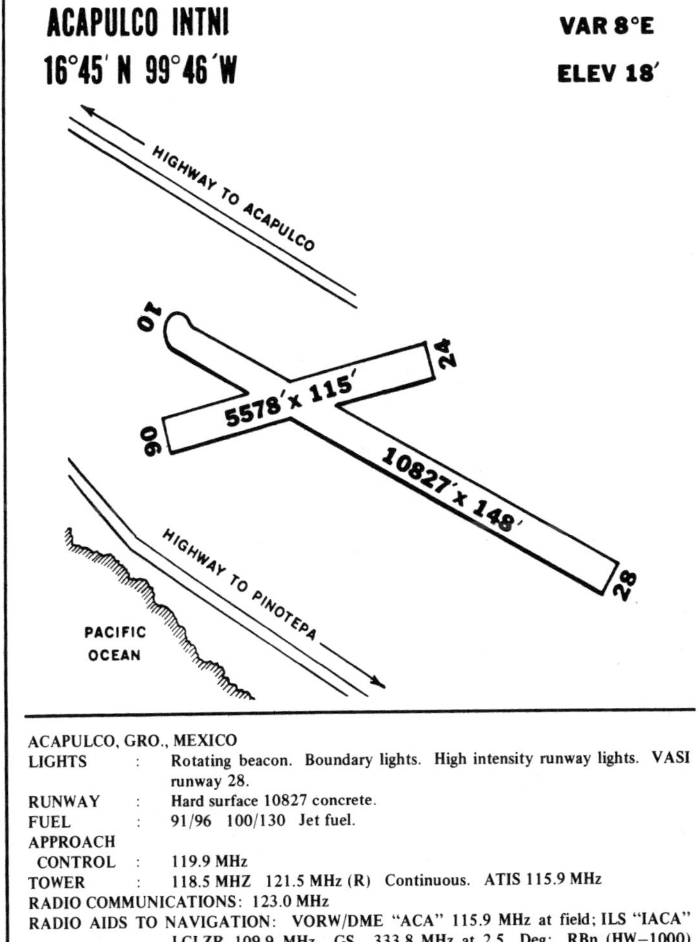

ACAPULCO, GRO., MEXICO

LIGHTS	:	Rotating beacon. Boundary lights. High intensity runway lights. VASI runway 28.
RUNWAY	:	Hard surface 10827 concrete.
FUEL	:	91/96 100/130 Jet fuel.
APPROACH CONTROL	:	119.9 MHz
TOWER	:	118.5 MHZ 121.5 MHz (R) Continuous. ATIS 115.9 MHz

RADIO COMMUNICATIONS: 123.0 MHz

RADIO AIDS TO NAVIGATION: VORW/DME "ACA" 115.9 MHz at field; ILS "IACA" LCLZR 109.9 MHz GS 333.8 MHz at 2.5 Deg: RBn (HW–1000) ACA 221 KHz at field. Continuous o pn on all NAVAIDS. LOM 350KHz Ident AC.

WEATHER	:	Hourly. Synoptic. Winds aloft.
REMARKS	:	Customs on request. Two-way radio required. VOR check point at intxn of taxiways ALFA & BRAVO, 091 RADIO FROM VOR.

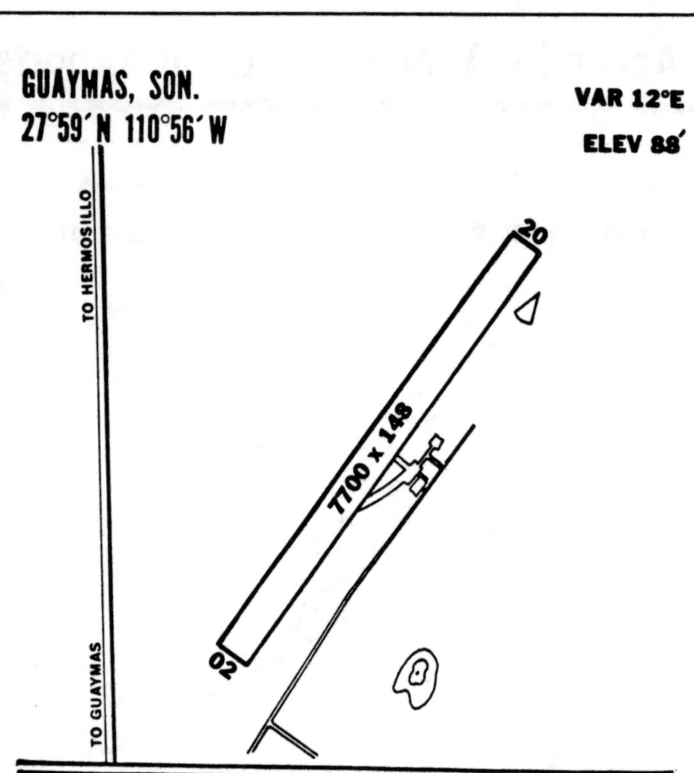

GUAYMAS, SON., MEXICO

LIGHTS	:	None.
RUNWAY	:	Hard surface 7700 ft.
FUEL	:	
TOWER	:	None

RADIO COMMUNICATIONS: (AFIS) 123.0 MHz 1200-2400Z.

NAVAIDS	:	NDB GYM 368 KHz approx 2 NM south of field continuous operation.
WEATHER	:	
REMARKS	:	Customs on request. Night operations not authorized. Preferential departure runway 02.

HERMOSILLO, LA MANGA

29°06′ N 111°03′ W

VAR 12° E

ELEV 615′

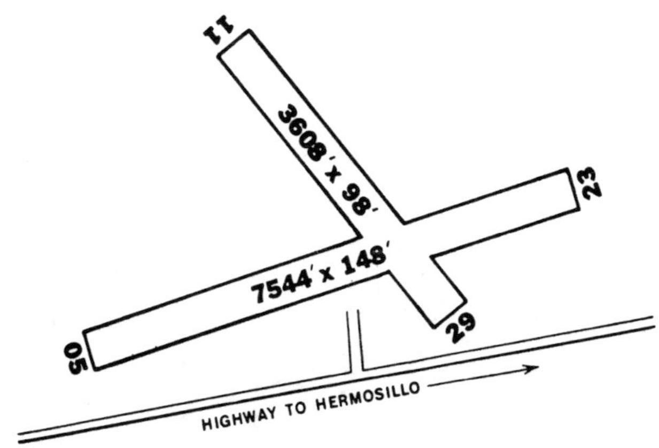

HERMOSILLO, SON., MEXICO

LIGHTS : Rotating beacon. High intensity lights runway 02-23. VASI runway 05. Taxiway lights.
RUNWAY : Hard surface 7544 asphalt.
FUEL : 80/87 100/130.
RADIO COMMUNICATIONS: 127.3 MHz 123.0 MHz (Hermosillo Radio) 1500-0300Z.
TOWER : 118.1 MHz 121.5 MHz (emergency) Continuous.
APPROACH CONTROL: 121.4 MHz 1300-0200Z
RADIO AIDS TO NAVIGATION: VORW/DME 112.8 MHz HMO at field. NAVAIDS continuous operation.
WEATHER : Hourly. Synoptic. Winds aloft.
REMARKS : Customs on request. Two-way radio required. Right traffic runways 05 and 11. Night operations not authorized runway 11-29.

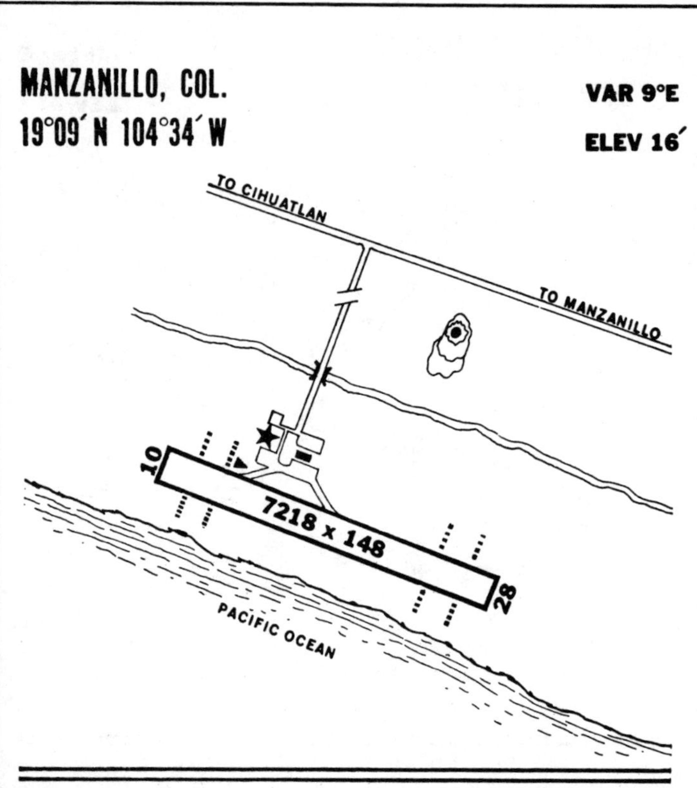

MANZANILLO, COL., MEXICO

LIGHTS	:	Airport beacon. High intensity runway lights runway 10 and 28. VASI runway 10 and 28.
RUNWAY	:	Hard surface 7218 ft.
FUEL	:	
TOWER	:	118.7 MHz 1400-0200Z.

RADIO COMMUNICATIONS: 123.0 MHz 1400-2400Z.

NAVAID	:	VOR/DME MZL 116.8 MHz west end of airport. Continuous operation.
WEATHER	:	Hourly. Synoptic.
REMARKS	:	Customs on request. Two-way radio required.

MAZATLAN LOS PATOS

VAR 10°E

23°09'N 106°15'W

ELEV 16'

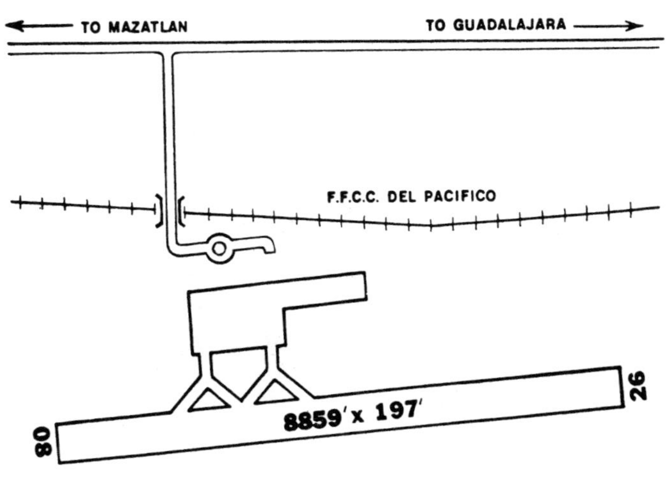

MAZATLAN, SIN., MEXICO

LIGHTS : Rotating beacon. High intensity lights runway 08-26. VASI runway 08 and 26. Lighted wind indicator.

RUNWAY : Hard surface 8859 asphalt.

FUEL : 80/87 91/96 100/130.

APPROACH CONTROL: 128.0 MHz 126.3 MHz 124.2 MHz 121.5 MHz Continuous service.

TOWER : 118.3 MHz 121.5 MHz (R) Continuous service ATIS 114.9 MHz.

RADIO COMMUNICATIONS: 123.0 MHz Mon/Sat 1200-0200Z Sun 1400-0200Z. 127.3 MHz 1300-0400 daily. 5568 KHz (D) 10017 KHz (D) 2966 KHz (N) 1300-0400 daily.

RADIO AIDS TO NAVIGATION: VORW/DME MZT 114.9 MHz at field continuous operation. RBn MZT 285 KHz 5.4 NM east approach end runway 26 1400-0400 daily.

WEATHER : Hourly. Synoptic. Winds aloft.

REMARKS : Customs on request. Two-way radio required.

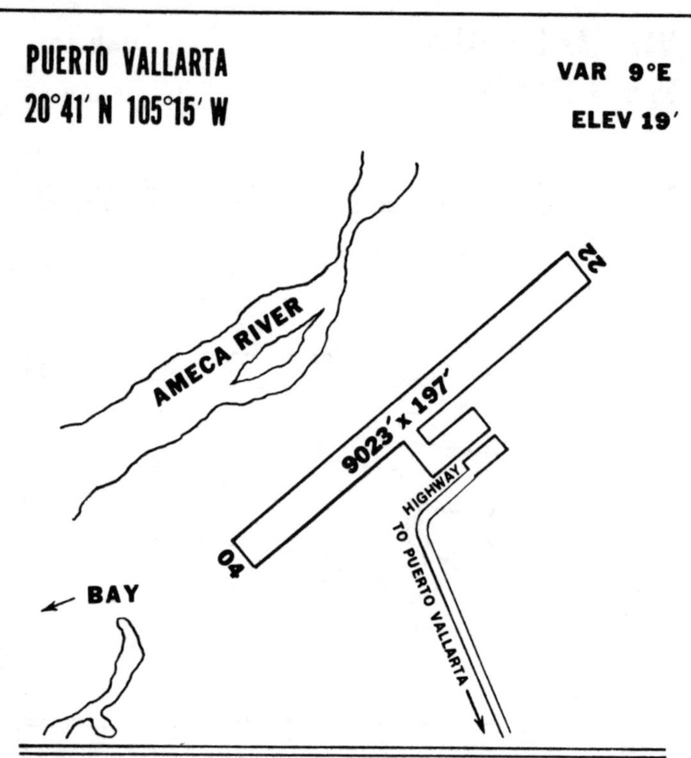

PUERTO VALLARTA, JAL., MEXICO

LIGHTS : High intensity runway lights. Taxiway lights. VASI runway 04. REIL runway 04.
RUNWAY : Hard surface 9023 asphalt.
FUEL : 80/87 100/130.
TOWER : 118.5 MHz 122.5 MHz (R) 1300-0200Z. 118.1 until July.
APPROACH CONTROL: 119.4 MHz below 20,000' 1330-0600Z daily.
RADIO COMMUNICATIONS: 127.3 MHz 123.0 MHz 1200-0400Z.
RADIO AIDS TO NAVIGATION: VORW/DME PVR 112.6 MHz at field, continuous opn. RBn PVR 304 KHz at field, continuous opn. ILS IPVR 109.5 MHz GS 332.6 MHz at 2.8 deg.
WEATHER : Hourly. Synoptic.
REMARKS : Two-way radio required. Customs on request. Runway 22 right traffic. VOR checkpoint at head runway 22 radio 029 from VOR.

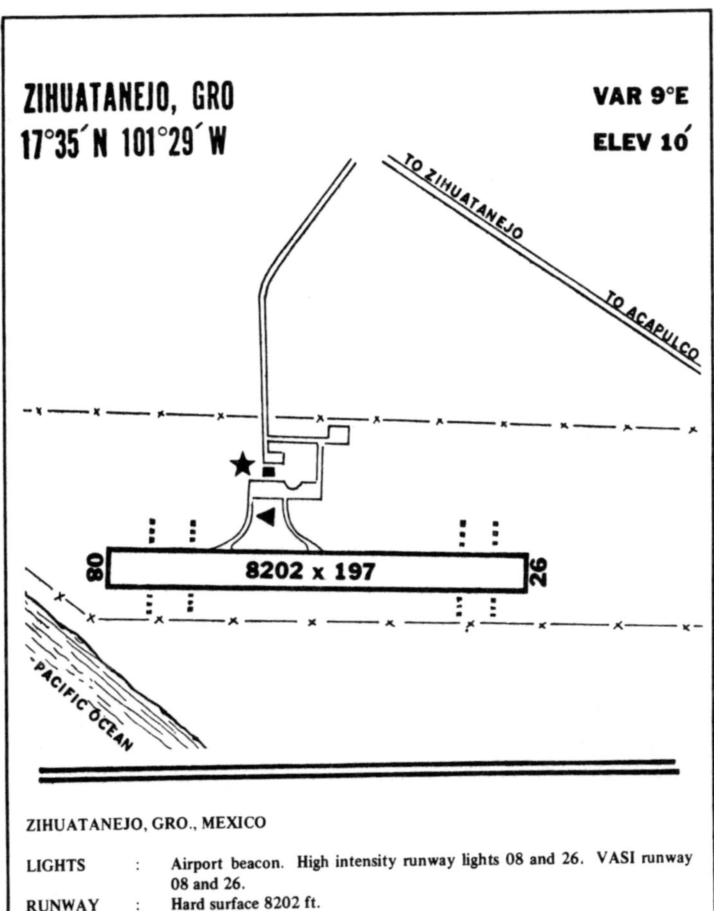

ZIHUATANEJO, GRO., MEXICO

LIGHTS	:	Airport beacon. High intensity runway lights 08 and 26. VASI runway 08 and 26.
RUNWAY	:	Hard surface 8202 ft.
FUEL	:	
TOWER	:	118.3 MHz 1300-0100Z.
RADIO COMMUNICATIONS:		123.0 MHz 1200-0000Z.
NAVAIDS	:	VOR/DME IH 113.8 MHz continuous operation.
WEATHER	:	Hourly. Synoptic.
REMARKS	:	Customs on request. Two-way radio required. Right traffic runway 08.

Appendix B: Useful Information

Appendix B: Useful Information

MEXICAN GOVERNMENT TOURISM DEPARTMENT OFFICES

Chicago, Illinois 60601
210 North Michigan Avenue

Dallas, Texas 75201
1905 Commerce Street

El Paso, Texas 79901
310 San Francisco Avenue

Houston, Texas 77002
Mellie Esperson Building 146

Los Angeles, California 90005
3106 Wilshire Boulevard

Miami, Florida 33131
First Nat'l Bank Bldg., Arcade, Office 20

New Orleans, Louisiana 70130
203 Saint Charles Street

New York, New York 10020
630 Fifth Avenue, Suite 3508

Phoenix, Arizona 85012
3443 North Central Avenue

San Antonio, Texas 78205
209 E. Travis Street
Travis Plaza Hotel

San Diego, California 92101
707 Broadway, Suite 935

Tucson, Arizona 85701
80 North Stone Avenue

San Francisco, California 94108
219 Sutter Street

Washington D. C. 20036
1028 Connecticut Avenue, Suite 606-B

MEXICAN COUNSULATES IN THE UNITED STATES

Albuquerque, New Mexico 87101
1306 Central Avenue, S.W.

Austin, Texas 78701
330 Perry Brooks Building

Boston, Massachusetts 02116
9 Newbury Street

Brownsville, Texas 78520
104-106 Majestic Building
1000 E. Elizabeth Street

Buffalo, New York 14202
610 Chamber of Commerce Building

Calexico, California 92231
307 Sherman Street

Chicago, Illinois 60606
Suite 842, 201 N. Wells

Cincinnati, Ohio 45202
Room 317, Provident Bank Bldg.
7th and Vine Streets

Corpus Christi, Texas 78401
601-602 Jones Building

Dallas, Texas 75201
527 Southland Center

Del Rio, Texas 78840
W. Greenwood and Main

Denver, Colorado 80202
410-412 Cochran Building
1031 15th Street

Detroit, Michigan 48201
1016-1017 Fox Building
2211 Woodward Avenue

Douglas, Arizona 85607
1053 F. Avenue

Eagle Pass, Texas 78852
391 Main Street

El Paso, Texas 79901
206 San Francisco Street

Fort Worth, Texas 76102
1113 W. T. Waggoner Building
810 Houston Street

Fresno, California 93721
305-308 Bank of America Bldg.
Fulton and Tulare Streets

Galveston, Texas 77550
3009 Avenue R 1/2

Honolulu, Hawaii 96805
P.O. Box 9098

Houston, Texas 76102
503 World Trade Building
1520 Texas Avenue

Indianapolis, Indiana 46208
4971 Olympia Drive

Kansas City, Missouri 64106
907 Walttower Building
East 9th. Walnut Street

Laredo, Texas 78040
1611 Farragut Street

Los Angeles, California 90013
354 S. Spring Street Apt. 179

Lubbock, Texas 79401
Suite 211
19th and Avenue M

McAllen, Texas 78501
519 S. Main Street

Marfa, Texas 79843
Paisano Hotel
P.O. Drawer M

Memphis, Tennessee 38105
410 N. Waldran Boulevard

Miami, Florida 33132
1205 Congress Building
111 NE 2nd Avenue

Milwaukee, Wisconsin 53202
229 E. Wisconsin Avenue

Mobile, Alabama 36602
59 North Water Street

Neenah, Wisconsin 54956
1106 E. Forest Street

New Orleans, Louisiana 70130
534 Whitney Bank Building

New York, New York 10017
8 East 41st Street

Newark, New Jersey 07102
48 Commerce Street

Nogales, Arizona 85621
135 Terrace Street

Oklahoma City, Oklahoma 73102
2916 First Nat'l Bank Building

Philadelphia, Pennsylvania 19107
Suite 2511, 12 S. 12th Street

Phoenix, Arizona 85014
14 N. Central Avenue

Pittsburgh, Pennsylvania 15219
306-308 Berger Building

Rochester, Minnesota 55901
3104 Harbor Heights Drive

Sacramento, California 95814
809 8th Street

St. Louis, Missouri 63101
510-512 Louderman Building
317 N. 11th Street

St. Paul, Minnesota 55107
284 E. Sidney Street

San Antonio, Texas 78205
127 Navarro Street

San Bernardino, Calif. 92401
392 W. Court Street
P.O. Box 245

San Diego, California 92101
625 Broadway

San Francisco, Calif. 94102
Suite 516, 870 Market Street

San Jose, California 95113
57 E. Santa Clara Avenue

San Juan, Puerto Rico
Calle Aibomito 1458

Seattle, Washington 98101
1904 3rd Avenue

Tampa, Florida 33606
516 Bay Street

Tucson, Arizona 85701
553 S. Stone Avenue
P.O. Box 2049

AMERICAN EMBASSY

American Embassy
Mexico Reforma 305
Mexico 5, D.F.

AMERICAN CONSULATES IN MEXICO

Location	*Mailing Address* *Mexico*	*U.S.*
American Consulate 16 de Septiembre 1896 Ciudad Juarez, Chihuahua	Apdo, Postal 164 Ciudad Juarez, Chih. Tel: 2-25-10 2-25-11	P.O. Box 10545 El Paso, Texas 79995
American Consulate General Progreso 175 Guadalajara, Jalisco	Apdo. Postal 1 Bis. Guadalajara, Jalisco Tel: 3-29-95/6/7/8 4-87-30/6/7/8/9	
American Consulate Ave Primera 232 between Azucenas and Azaleas Matamoros, Tamaulipas	Apartado Postal 451 Matamoros, Tamaulipas Tel: 2-02-41	P.O. Box 633 Brownsville, Texas 78520
American Consulate Edificio la Siesta Hotel Esquina Olas Altas y Mariano Escobedo Mazatlán, Sinaloa	Apdo. Postal 321 Mazatlan, Sinaloa Tel: 26-85 26-87 29-05	

AMERICAN CONSULATES IN MEXICO (CONTINUED)

Location	*Mailing Address* *Mexico*	*U.S.*
American Consulate Paseo Montejo 453 Merida, Yucatan	Apartado Postal 130 Merida, Yucatan Tel: 1-26-03 1-60-30	
American Consulate Avenida Arista 1974 Mexicali, Baja Calif.	Apdo. Postal 402 Mexicali, Baja Calif. Tel: 762-6312/3 762-6303	P.O. Box 1192 Calexico, California 92231
American Consulate General Edificio del Roble Avenida Juarez 800 Sur Monterrey, Nuevo Leon	Apdo. Postal 152 Monterrey, Nuevo Leon Tel: 43-06-50	
American Consulate Edif. Pronap 3 Piso Avenidas Sonora y Lopez Mateos Nogales, Sonora	Edif. Pronap 3 Piso Avenidas Sonora y Lopez Mateos Nogales, Sonora Tel: None	P.O. Box 1090 Nogales, Ariz. 85621
American Consulate Calle Madero y Avenida Ocampo Nuevo Laredo, Tamaulipas	Apdo. Postal 38 Nuevo Laredo, Tamaulipas Tel: 2-00-05	P.O. Box 449 Laredo, Texas 78040
American Consulate Padre de Las Casas y Allende Piedras Negras, Coahuila	Apdo. Postal 20 Piedras Negras, Coahuila Tel: 34	P.O. Box AA Eagle Pass, Texas 78852
American Consulate Edificio Plaza Diaz Miron 106, Oriente Tampico, Tamaulipas	Apdo. Postal 100 Tampico, Tamaulipas Tel: 2-36-00/01	
American Consulate General Tapachula 96 Tijuana, Baja Calif.	Apdo. Postal 68 Tijuana, Baja Calif. Tel: 386-1001/2/3/4/5 386-3003/4/5/6	P.O. Box 1358 San Ysidro, California 92073
American Consulate Arista y Malecon Veracruz, Veracruz	Arista y Malecon Veracruz, Veracruz Tel: 2-30-40 2-22-53	

AMERICAN CONSULATES IN MEXICO (CONTINUED)

Location	*Mexico*
American Consulate General Blvd. Miguel Nidalgo 15 Edif. Isstason 3 Piso Hermosillo, Sonora	Apdo. Postal 972 Hermosillo, Sonora Tel: 3-89-23 3-89-24 3-89-25
American Consulate Portal Matamoros 98 Morelia, Michoacan	Apdo. Postal 151 Morelia, Michoacan Tel: 2-04-66
American Consulate Avenida Carranza 870 San Luis Potosi, S.L.P.	Apdo. Postal 697 San Luis Potosi, S.L.P. Tel: 2-13-30

SAMPLE MEXICAN AVIATION WEATHER REPORT

```
MEX   1800Z   RS2   -XE30⦶/⊕ 13/10TRW-H   110/24/19NE15+35/
 1      2      3       4       5   6   7   8  9          10

141/T
 11

SBR STN MOV SE RLPG NUBE/TIERRA H3/60324/90321/13
               12                            13     14   15
```

BREAKDOWN OF REPORT:

1. Station identification (MEX). Mexico City.

2. Time of report (1800Z). Greenwich Mean Time.

3. Type of report (RS2). Second special report originated by Mexico City for the day.

   R = Regular
   S = Special
   RS = Regular Special
   E = Extra
   V = Verification

4. Sky condition (-XE30 ⦶/⊕) Sky partially obscured, estimated ceiling three thousand broken, overcast cirroform.

   Sky and ceiling conditions are coded the same as in U.S. Weather Bureau Manual of Surface Observations, Circular N.

5. Visibility (13-10). One and three-tenths nautical miles.

The visibility is reported in nautical miles and tenths of nautical miles.

6. Weather (TRW-). Moderate thunderstorm and light rainshowers.

Coded symbols and intensity same as in U.S. Weather Bureau Manual of Surface Observations, Circular N.

7. Obstructions to visibility (H). Haze.

Obstructions to visibility are coded the same as in U.S. Weather Bureau Manual of Surface Observations, Circular N.

8. Atmospheric pressure or sea level pressure (110). 1011.0 Millibars.

This data is only included in synoptic reports at three-hour intervals.

9. Ambient and dewpoint temperature (24/19). Reported in whole degrees centigrade.

10. Wind direction and velocity (NE15+35). Wind direction is reported to 16 degrees of the compass and velocity is reported in knots.

11. Altimeter setting (141). 1014.1 millibars. Reported in tens, units and tenths of millibars.

12. Remarks (T SBR STN MOV SE RLPG NUBE/TIERRA H3). Thunderstorm over station, moving southeast, cloud to ground lightning and haze in obstructions to visibility is obscuring 3/10 of the sky.

13. Coded remarks (60324). Coded barometric tendency and pressure change in past three hours and total amount of rainfall in hundredths of inches in past six hours.

14. Cloud group (90321). Coded group indicating type of low, middle, and high clouds, and direction of movement of clouds.

15. Maximum or minimum temperature (13). Minimum temperature 13 degrees centigrade past 24 hours.

Index

Edited by Steven Mesner